David Berlinski

A Tour of the Calculus

David Berlinski was born in New York City. He received
a B.A. degree from Columbia College and a Ph.D. from
Princeton University. Having a tendency to lose academic
positions with what he himself describes as an embar-
rassing urgency, Berlinski now devotes himself entirely to
writing. He lives in San Francisco.

A
Tour
of
the
Calculus

A
Tour
of
the
Calculus

David Berlinski

VINTAGE BOOKS

A DIVISION OF RANDOM HOUSE, INC.

NEW YORK

Grateful acknowledgment is made to the following for permission to
reprint previously published material:

Dutton Signet: Excerpt from "Of Exactitude in Science" from *A Universal
History of Infamy* by Jorge Luis Borges, translated by Norman Thomas di
Giovanni, translation copyright © 1970, 1971, 1972 by Emece Editores,
S.A., and Norman Thomas di Giovanni. Reprinted by permission of
Dutton Signet, a division of Penguin Books USA Inc.

Alfred A. Knopf, Inc.: Excerpt from "Thirteen Ways of Looking at a
Blackbird" from *Collected Poems* by Wallace Stevens, copyright © 1954 by
Wallace Stevens. Reprinted by permission of Alfred A. Knopf, Inc.

The Library of Congress has cataloged the Pantheon edition as follows:
Berlinski, David.
A tour of the calculus / David Berlinski.
p.   cm.
ISBN 0-679-42645-0
1. Calculus—Popular works. I. Title.
QA303.B488 1995
515—dc20   95-4042
Vintage ISBN: 0-679-74788-5

Book design by Deborah Kerner

Random House Web address: http://www.randomhouse.com/

Printed in the United States of America
79B86

For my Victoria

Long live the sun. May the darkness be hidden.

contents

introduction

As its campfires glow against the dark, every culture tells stories to it-self about how the gods lit up the morning sky and set the wheel of being into motion. The great scientific culture of the West—*our* cul-ture—is no exception. The calculus is the story this world first told it-self as it became the modern world.

The sense of intellectual discomfort by which the calculus was pro-voked into consciousness in the seventeenth century lies deep within memory. It arises from an unsettling contrast, a division of experience. Words and numbers are, like the human beings that employ them, iso-lated and discrete; but the slow and measured movement of the stars across the night sky, the rising and the setting of the sun, the great ball bursting and then unaccountably subsiding, the thoughts and emotions that arise at the far end of consciousness, linger for moments or for months, and then, like barges moving on some sullen river, silently dis-appear—these are, all of them, continuous and smoothly flowing processes. Their parts are inseparable. How can language account for what is not discrete, and numbers for what is not divisible?

Space and time are the great imponderables of human experience, the continuum within which every life is lived and every river flows. In its largest, its most architectural aspect, the calculus is a great, even spectacular theory of space and time, a demonstration that in the real numbers there is an instrument adequate to their representation. If sci-ence begins in awe as the eye extends itself throughout the cold of space, past the girdle of Orion and past the galaxies pinwheeling on their axes, then in the calculus mankind has created an instrument commensurate with its capacity to wonder.

It is sometimes said and said sometimes by mathematicians that the usefulness of the calculus resides in its applications. This is an incoherent, if innocent, view of things. However much the mathematician may figure in myth, absently applying stray symbols to an alien physical world, mathematical theories apply *only* to mathematical facts, and mathematics can no more be applied to facts that are *not* mathematical than shapes may be applied to liquids. If the calculus comes to vibrant life in celestial mechanics, as it surely does, then this is evidence that the stars in the sheltering sky have a secret mathematical identity, an aspect of themselves that like some tremulous night flower they reveal only when the mathematician whispers. It is in the world of things and places, times and troubles and dense turbid processes, that mathematics is not so much applied as *illustrated*.

Whatever physicists may say, both space and time, it would seem, go on and on; the imaginary eye pushed to the very edge of space and time finds nothing to stop it from pushing further, every conceivable limit a seductive invitation to examine the back side of the beyond. We are finite creatures, bound to this place and this time, and helpless before an endless expanse. It is within the calculus that for the first time the infinite is charmed into compliance, its luxuriance subordinated to the harsh concept of a limit. The *here* and *now* of ordinary life, these are coordinated by means of a mathematical function, one of the noble but inscrutable creations of the imagination, the silken thread that binds together the vagrant world's far-flung concepts. Fabulous formulas bring anarchic speed panting to heel and make of its forward rush a function of time; the wayward area underneath a curve is in the end subordinated to the rule of number. Speed and area, the calculus reveals, are related, the revelation acting like lightning flashing between two distant mountain peaks, the tremendous flash of light showing in the moment before it subsides that those peaks are strangely symmetrical, each existing to sustain the other. The relationship that holds between speed and area holds also between concepts that are *like* speed and area, the calculus emerging at the far end of these considerations as the most general of theories treating continuous magnitudes, its concepts appearing in a thousand scattered sciences, the light that they shed *there* reflections of the subject's central light, its single sun.

The dryness of this description should not obscure the drama that it reveals. Of all the miracles available for inspection, none is more striking than the fact that the real world may be understood in terms of the real numbers, time and space and flesh and blood and dense primitive throbbings sustained somehow and brought to life by a network of secret mathematical nerves, the juxtaposition of the two, throbbings on the one hand, those numbers on the other, unsuspected and utterly surprising, almost as if some somber mechanical puppet proved capable of articulated animation by means of a distant sneeze or sigh.

The body of mathematics to which the calculus gives rise embodies a certain swashbuckling style of thinking, at once bold and dramatic, given over to large intellectual gestures and indifferent, in large measure, to any very detailed description of the world. It is a style that has shaped the physical but not the biological sciences, and its success in Newtonian mechanics, general relativity, and quantum mechanics is among the miracles of mankind. But the era in thought that the calculus made possible is coming to an end. Everyone feels that this is so, and everyone is right. Science will, no doubt, continue as a way of life, one among others, but its unique claim to our intellectual or religious devotion—this has been lost and it is foolish to deny it.

This is an elegiac conclusion. Although what I have written is intended to evoke a miracle, in the end there is that melancholy fall, a reminder that in mathematics as in everything else, stories have a beginning and they come to an end.

SAN FRANCISCO, 1995

a note to the reader

The fundamental theorem of the calculus is the focal point of this book, the goal toward which the various chapters tend. The book has a strong narrative drive, its various parts subordinated to the goal of enabling anyone who has read what I have written to experience that hot flush that accompanies any act of understanding, saying as he or she puts down the book, *Yes, that's it, now I understand.*

My aim throughout has been to provide a tour of the calculus, not a treatise. I have concentrated on the essentials. There are no problem sets or exercises or anything at all like that in what I have written. I have suppressed whenever possible mathematical formalism in favor of ordinary English. But there is no explaining mathematics without from time to time using mathematics, and the mathematician's symbolism, which to an outsider looks as inviting as Chinese, *does* represent an instrument of matchless power and concision. It is my hope that by using this instrument sparingly the symbols might come to gleam against the background of plain prose, like jewels seen on black velvet.

The argument of the book is conveyed in the text itself; down below, in the various appendices, definitions are given in their full formality and a number of theorems demonstrated. I have not proved everything that might be proved: some statements remain in the text as ringing affirmations. Nothing in the appendices is beyond the grasp of the ordinary reader, but there is no avoiding the fact that confrontation with proof is quite often a humbling experience. The eye slows; a feeling of helplessness steals over the soul. At first, it seems as if the confident language of mathematical assertion constitutes a subtle form of mockery. There is no help for any of this save the ancient remedies of practice and a willingness to put pencil to paper. Readers who want the big picture need not linger in the cellars; but a mathematical argument, once understood, is in its capacity to compel belief a miracle of enlightened life. Those who at first recoil indignantly from a disciplined argument may in time revisit appreciatively the inferences they rejected.

I have written this book for men and women who wish to understand the calculus as an achievement in human thought. It will not make them mathematicians, but I suspect that what they want is simply a little more light shed on a dark subject.

And that is something we all could use: a little more light.

the frame of the book

The overall structure of the calculus is simple. The subject is defined by a fantastic leading idea, one basic axiom, a calm and profound intellectual invention, a deep property, two crucial definitions, one ancillary definition, one major theorem, and the fundamental theorem of the calculus.

*The **fantastic leading idea:** the real world may be understood in terms of the real numbers.*

*The **basic axiom:** brings the real numbers into existence.*

*The **calm and profound invention:** the mathematical function.*

*The **deep property:** continuity.*

*The **crucial definitions:** instantaneous speed and the area underneath a curve.*

*The **ancillary definition:** a limit.*

*The **major theorem:** the mean value theorem.*

*The **fundamental theorem** of the calculus is the fundamental theorem of the calculus.*

These are the massive load-bearing walls and buttresses of the subject.

A
Tour
of
the
Calculus

chapter 1

Masters
of the
Symbols

SOME THINGS WERE GREEK TO THE GREEKS. IN THE FIFTH CENTURY B.C., Zeno the Eleatic argued that a man could never cross a room to bump his nose into the wall.

How so?

In order to reach the wall he would have first to cross half the room, and then half the remaining distance again, and then half the distance that yet remains. "This process," Zeno wrote in an argument still current in fraternity houses (where it never fails to impress the brothers), "can always be continued and can never be ended." But an infinite process requires an infinite amount of time for its completion, no? So one might think. A process is, after all, something that takes place in time. But the plain fact is that we *are* capable of compressing those infinite steps into a brisk walk from one end of a room to the other: *that* sort of thing we do with ease. An irresistible inference is in conflict with an inescapable fact, Zeno's diamond-bright little argument serving to invest the ordinary with a lurid aspect of the impossible.

It is now twenty-two centuries later, time pivoting at the seventeenth century to pause reflectively and answer Zeno. No telephones yet; no fax machines; no cappuccino; no computers; no roads really. Stairs but no StairMasters. Sanitation? Appalling. Ditto for personal hygiene. But no MTV either, no late-night infomercials for wok cooking or Swedish hair restoratives, Madonna an incubus merely, waiting to be born. Before the seventeenth century, everything is squid ink and ocean ooze and dark clotted intuitions; but afterward, a strange symbolic system erupts into existence and floods the intellectual landscape with a hard flat nacreous light. Communing with the powers of the night and the dark undulating rhythms that flow across the sky, the mathematician—of all people!—emerges as the unexpected master of those symbols, the calculus his treasure chest of chants and incantations, fabulous formulas, wormholes into the forbidden heart of things.

In its historical development, the calculus represents an exercise in delayed gratification. *Gratification* is the right if unexpected word, suggesting as it does a moist intellectual explosion, but *delay* is the governing concept, the calculus like one of those poignant adolescent dreams in which desires are painfully defined but hopelessly deferred. The warm-up to the calculus stretches from the ancient world to the seventeenth century, but the subject's center was discovered quite suddenly by Gottfried Leibnitz and Isaac Newton in the second half of the century, a striking example of two fire alarms going off in the night at precisely the same time in two widely separated countries. Other mathematicians, in France, England, and Italy, it is true, saw this and they saw that, but they never saw this *and* that and so remain in history forever holding the door through which Leibnitz and Newton raced. Like every story, this one has a before and an after, a passage from darkness into light. The common view is the commonsensical view: in Leibnitz and Newton there was an effulgence, a shining forth.

It goes without saying, of course, that although each man had plainly conceived his ideas uninfluenced by the other, both wasted enormous energies in an undignified and peevish effort to establish the priority of their claims.

Isaac Newton was born on Christmas Day, 1642; it was the year that Galileo died. A curious series of numerical coincidences runs through the history of the calculus. Early portraits show him as a saturnine youth, with a long face marked by a high forehead and small suspicious eyes. It is not the face of a man inclined much to small talk or to pleasant evenings spent in steamy pubs, a glass of bitter in hand. The tension at his mouth suggests someone prepared to withdraw quivering in irritation from his senses. And those small, sharp, shrewd but dark and narrow eyes, they seem to say, those eyes: *Let me see now, Mr. Berlinski, your deductions for this year appear to exceed your income* . . .

It is *that* sort of face.

In the winter of 1665–66, Trinity College in Cambridge closed its doors owing to the plague. Newton returned to his home in the English countryside. At twenty-three he was already marked by his contemporaries as a man with a deep indwelling nature, an indifference to pleasure. In the year that followed, Newton stated and proved the binomial theorem (a generalization of the familiar rule that $a + b$ times itself is $a^2 + 2ab + b^2$), invented the calculus, discovered the universal law of gravitation (and so created contemporary dynamics), and developed a theory of color. In English history, those twelve months are known quite properly as the *annus mirabilis,* the year of miracles. On returning to Cambridge, Newton was appointed the Lucasian professor of mathematics. He made his discoveries known with the natural reluctance of a man convinced of his genius and so indifferent to praise, but in 1687, at the urging of the astronomer Edmund Halley, he published the *Philosophiae Naturalis Principia Mathematica,* commonly known as the *Principia,* and thereby secured his enduring reputation as the author of the greatest scientific work in history.

The *Principia* is the supreme expression in human thought of the mind's ability to hold the universe fixed as an object of contemplation; it is difficult to reconcile its monumental power with a number of humanly engaging but anecdotal accounts of its composition: the disheveled and half-dressed Newton, so the stories run, his crumb-filled wig askew, shambling about the evil-smelling room in which he lived and worked, muttering to himself, his thin lips half forming words, stiff with attention or slack and slumped indifferently on his unmade bed, entirely absorbed, forgetting to eat and sleeping in weak, disorganized fits, an apple rotting on the desk, the *Principia* taking shape in stages, vellum sheets piling up on the wooden desk.

It is the place where modern physics begins, this vatic text, and so in a certain sense the place where modern *life* begins. Stars in the staring sky *and* objects on the surface of earth are in the *Principia* brought under the control of a simple symbolic system, their behavior circumscribed by the law of universal attraction. The anarchic way-wardness of the pre-Newtonian universe is gone for good, the gods who had gone before scattered to the night winds in favor of the red-eyed God that for a time Newton alone could see. The universe in *all of its aspects,* the *Principia* goes on to suggest, is coordinated by a Great Plan, an elaborate and densely reticulated set of mathematical laws, a system of symbols. It is this idea that drove Newton. A portion of the *Principia* he majestically entitled *the system of the world*. It is this idea that yet drives physicists. Searching for a final theory, one that would subsume all other physical theories, Steven Weinberg, he of the Nobel Prize, is a Newton legatee, an heir.

And here is Gottfried Wilhelm Leibnitz, born in Leipzig just four years after Newton. He is standing by the *hors d'oeuvres* and the pot-ted shrimp, a fleshy man of perhaps forty. An enormous brunette wig with elaborate curls covers his head; he is dressed for court in lace and silk. He has a high forehead, arched cheekbones, wide-set staring eyes, and a large handsome nose; his is the face of a man, I think, who would enjoy mulled wine, poached eggs on buttered toast, a warm fire as the wind rattles the windows of a country castle, a young serving girl bending low over the plates and after dinner saying softly but without real surprise: *Why, Herr Leibnitz, really now, bitte!*

Leibnitz studied law, theology, and philosophy; he was interested in mathematics and diplomacy, history, geology, linguistics, biology, numismatics, classical languages, and candlemaking. A serenely con-fident man of high intellectual power with a steady, easily sustained interest in things, he spent much of his life in the services of the Hanoverian court in Germany, attending to weedy dukes wise enough to know their better when they met him. As a court official, he immersed himself in genealogy and legal affairs, traveling the Continent at the behest of his royal masters; but no matter his official duties, or the endless days spent cramped in wooden coaches, the bumpy roads of Europe beneath his well upholstered backside, he remained a metaphysician among metaphysicians and a mathematician among mathematicians—a Prince among Princes; he knew the great intellects of Europe and the great intellects of Europe knew him.

The contrast to Newton is instructive. Leibnitz was an intellectual man about town, what the French call *un brasseur d'affaires,* someone who saunters through a world of ideas; he came to the calculus because his genius caught on something and then *gushed.* The vision that he embraced was intensely *local.* There are problems here, things to study there, a world of overflowing variety. The simple rules governing affairs in Leipzig are not the intricate and complicated rules needed to make sense of sinister intrigues in Paris. The night is different from the day, the earth from the moon. What is appropriate in a *stuberl* is inappropriate at court. The sensible intelligence requires not so much universal laws as universal *methods,* ways of coordinating information and holding different aspects of the world together simultaneously. Leibnitz prophetically imagined a universal computing machine; he conceived the idea of a formal system; he understood, or so it appears, the nature of those discrete combinatorial systems that inform both human grammars and DNA; he saw in the future the shape that mathematical logic would take, and in his strange philosophical invocations of items such as monads, each of which somehow contains a potential universe, he seemed to divine the future course of quantum mechanics and cosmology, almost as if amidst the disorder and distractions of his life he was occasionally able to slip sideways into the stream of time and see just enough of the future to suggest his most pregnant and compelling ideas.

Newton, on the other hand, was an intellectual seer. The same hypnotic, coal-black eyes peer out intently from every mask he wore. He was driven to invent the calculus because it was the indispensable mathematical tool without which he could not complete—he could not *begin*—the enterprise involved in describing the Great Plan in all its limpidness, simplicity, and unearthly beauty. His vision of things was intensely *global*. The world's ornamental variety he regarded as an impediment to understanding. Nothing in his temperament longed to cherish the particular—the way in which wisteria smells in spring, the slow curve of a river bed, a woman's soft and puzzled smile, the overwhelming *thisness* of this or the *thatness* of that. Whatever the differences between one place and another, or between the past, the present, and the future, some underlying principle, some form of unity, subsumes them, those differences, and shows, to the mathematician at least, that like the cut edges of a glimmering crystal they are superficial aspects of a central flame.

This may suggest that between Leibnitz and Newton there was a difference in intellectual depth. Not so. I am talking of men of genius. And yet there is no doubt that it has been Newton's vision of the universe coordinated by a Great Plan, a set of mathematical principles pregnant enough to *compel* the very foundations of the world into being, that has until now been impressed on the physical sciences, so that the very enterprise itself, from the *Principia* to various theories of absolutely everything that contemporary physicists assure us are in preparation, bears the stamp of his enigmatic and brooding personality.

chapter 2

Symbols
of the
Masters

IT IS A FACT. AT SOME TIME OR OTHER THE MATHEMATICIANS OF EUROPE looked out over the universe, noted its appalling clutter, and determined that on some level there must exist a simple representation of the world, one that could be coordinated with a world of numbers. Note the double demand. A *representation* of the world, and one coordinated with *numbers*. When did this fantastic idea come about? I have no idea. It did not occur to the ancients, however much they may have been given to number mysticism; cowled and hooded medieval monks would have regarded the idea as superstitious mummery (as perhaps it is); and as late as the middle of the sixteenth century, amidst a culture that had learned brilliantly to represent aurochs and angels in terms of paint and durable pigment, the idea of a mathematical representation of the world remained alien and abstract. But by the end of the seventeenth century, the representation was essentially complete (even though it required another one hundred and fifty years for the logical details painfully to be put in place). The real world had been reinterpreted in terms of the real numbers. This fantastic achievement is the expression of a great psychological change, the moment of its completion comparable to the measured minute in antiquity during which the hectoring and complaining gods of the ancient world came to be seen as *aspects* of a single inscrutable and commanding deity.

The idea that the world at large (and so the world of experience) requires a mathematical representation raises two obvious questions. *Which* world is to be coordinated with numbers? And coordinated with *which* numbers? First things first. The mathematical representation of the world proceeds by means of Euclidean geometry, a theory old already in the seventeenth century. A vexed pause now to recollect high-school geometry. There is Mrs. Crabtree, standing glumly by the blackboard. There is Amy Kranz, dressed in a red sweater, her pubescent back arched invigoratingly. There is Stokely, the class clown, wadding up a spitball. But what is going on? In class, I mean. Apparently something to do with triangles or trapezoids. The blackboard is filled with drawings. And from purely an intuitive point of view, this snapshot (from the blessed fifties, in my own case) will do as well as anything else. Elementary geometry is the study of certain simple, regular, and evident shapes. Straight lines and points predominate. Except for a few simple arcs, no curves beyond the circle. No crooked lines. Nothing by way of irregularity or shapelessness. No algebra. Few symbols, in fact. The discipline proceeds by elimination and idealization. The meaty players are stripped from the muddy football field and the field itself reduced to its essentials of length, width, and area.

In its historical aspect, geometry is a subject that rises steaming from ancient Egyptian marshes, where tough overseers wearing oiled braids looked out over the fields, a stiff papyrus sheet underneath their arm, with even the most unapproachable of ancient rulers, The King Whose Name None Dare Speak, deferring to the man capable of determining the area under His cultivation or the volume of His awful pyramid. To recall that overseer is to recall the practical origins of the subject. Geometry as a high intellectual art leaves the overseer knee-deep in marsh and mud, a mosquito buzzing fitfully over his bronzed and polished head. The Greeks of the third century B.C., to whom the subject is due, took the overseer's lore and made of it a deductive science. Certain geometrical assertions were set aside and simply accepted as self-evident. A straight line, Euclid buoyantly affirmed, may be drawn between any two points. And then again, he affirmed again, all right angles are equal. There are five such postulates in Euclidean geometry, and a number of auxiliary axioms dealing with purely logical matters—the familiar declaration, for example, that equals added to equals are equal. From these postulates and axioms, Euclid proceeded to *derive* the assertions of geometry, its central theorems. He thus gave to the overseer's lore an enduring intellectual structure.

For many centuries the austere edifice of Euclidean geometry stood as a supreme example of pure thought. Euclid, it was said (by Edna St. Vincent Millay, who knew no geometry), looked on beauty bare. Its intellectual grandeur aside, Euclidean geometry plays a simple striking role in the organization of experience. It is a schematic; it functions as a blueprint. In Euclidean geometry, the outlines of the Great Plan are for the first time revealed. The straightforward definitions and theorems of Euclidean geometry, conceived initially as exercises in thought, the mind companionably addressing itself, have a direct and thus an uncanny interpretation in the voluptuous and confusing world of the senses. A straight line *is* the shortest distance between two points. That the structure of the physical universe seems to have been composed with Fitzwater and Blutford's high-school textbook, *Welcome to Geometry,* firmly in mind is evidence that in general things are stranger than they seem.

Humped, ancient, and austere, Euclidean geometry is a static theory and thus to some degree a stagnant theory; within its confines, everything remains the same, and from its lucid mirror no form of change is ever shown. Things are what they are, now and forever. This was a view favored by the Greeks who took the long view, indeed, of things; but *we* live in a world of ceaseless growth and decay, with things in fretful motion on the surface of the earth, planets wheeling in the night sky, galaxies coming into existence and then disappearing, and even the universe itself arising out of a preposterous *Bang!* and thus fated one day either to expand infinitely into the void or collapse back onto itself like a crushed Mallomar. Geometry may well describe the skeleton, but the calculus is a living theory and so requires flesh and blood and a dense network of nerves.

Adieu, Mrs. Crabtree, *adieu*.

How Much and How Many

Unlike Euclidean geometry, arithmetic rises directly from the wayward human heart, the lub-dub under the physician's stethoscope or the lover's ear (sounding very much like the words *so soon, it ends*), impossible to hear without a mournful mental echo: 1, 2, 3, 4, . . . , the doubled sounds, that beating heart, those numerical echoes, cohering perfectly for as long as any of us can count.

The most familiar of objects, numbers are nonetheless surprisingly slippery, their sheer slipperiness interesting evidence that certain intellectual tools may be successfully used before they are successfully understood. Numbers tend to sort themselves out by clans or systems, with each new system arising as the result of a perceived infirmity in the one that precedes it. The natural numbers 1, 2, 3, 4, . . . , start briskly at 1 and then go on forever, although how we might explain what it means for anything to go on forever without in turn using the natural numbers is something of a mystery. In almost every respect, they are, those numbers, simply given to us, and they express a primitive and intimate part of our experience. Like so many gifts, they come covered with a cloud. Addition makes perfect sense within the natural numbers; so, too, multiplication. Any two natural numbers may be added, any two multiplied. But subtraction and division are curiously disabled operations. It is possible to subtract 5 from 10. The result is 5. What of 10 from 5? No answer is forthcoming from *within* the natural numbers. They *start* at 1.

The integers represent an expansion, a studied enlargement, of the system of natural numbers, one motivated by obvious intellectual distress and one made possible by two fantastic inventions. The distress, I have just described. And those inventions? The first is the number 0, the creation of some nameless but commanding Indian mathematician. When 5 is taken away from 5, the result is nothing whatsoever, the apples on the table vanishing from the table, leaving in their place a peculiar and somewhat perfumed absence. What *was* there? Five apples. What *is* there? Nothing, *Nada,* Zip. It required an act of profound intellectual audacity to assign a name and hence a symbol to all that nothingness. Nothing, *Nada,* Zip, Zero, 0.

The negative numbers are the second of the great inventions. These are numbers marked with a caul: −504, −323, −32, −1 (I have always thought the minus sign a symbol of strangeness). The result is a system that is centered at 0 and that proceeds toward infinity in both directions: . . . , −5, −4, −3, −2, −1, 0, 1, 2, 3, 4, ... Subtraction is now enabled. The result of taking 10 from 5 is −5.

And yet if subtraction (along with addition and multiplication) is enabled among the integers, division still provokes a puzzle. Some divisions may be expressed entirely in integral terms—12 divided by 4, for example, which is simply 3. But what of 12 divided by 7? Which in terms of the integers is nothing whatsoever and so calls to mind those moments on *Star Trek* when the transporter fails and causes the Silurian ambassador to vanish.

It is thus that the rational numbers, or fractions, enter the scene, numbers with a familiar doubled form: 2/3, 5/9, 17/32. The fractions express the relationship between the whole of things that have parts and the parts that those things have. There is that peach pie, the luscious whole, and there are those golden dripping slices, parts of the whole, and so two thirds or five ninths or seventeen thirty-seconds of the thing itself. With fractions in place, division among the integers proceeds apace. Dividing 12 by 7 yields the exotic 12/7, a number that does not exist (and could not survive) amidst the integers. But fractions play in addition a conspicuous role in measurement and so achieve a usefulness that goes beyond division.

The natural numbers answer the oldest and most primitive of questions—*how many?* It is with the appearance of this question in human history that the world is subjected for the first time to a form of conceptual segregation. To count is to classify, and to classify is to notice and then separate, things falling within their boundaries and boundaries serving to keep one thing distinct from another. The world before the appearance of the natural numbers must have had something of the aspect of an old-fashioned Turkish steambath, pale, pudgy figures arising out of the mist and shambling off down indistinct corridors, everything vague and vaguely dripping; afterward, the world becomes hard-edged and various, the discovery of counting leading ineluctably to an explosive multiplication of bright ontological items, things newly created because newly counted.

The rational numbers, on the other hand, answer a more modern and sophisticated question—*how much?* Counting is an all or nothing affair. Either there are three dishes on the table, three sniffling patients in the waiting room, three aspects to the deity, or there are not. The question *how many?* does not admit of refinement. But *how much?* prompts a request for measurement, as in *how much does it weigh?* In measurement some extensive quantity is assessed by means of a scheme that may be made better and better, with even the impassive and uncomplaining bathroom scale admitting of refinement, pounds passing over to half pounds and half pounds to quarter pounds, the whole system capable of being *forever* refined were it not for the practical difficulty of reading through the hot haze of frustrated tears the awful news down there beneath all that blubber. This refinement, which is an essential part of measurement, plainly requires the rational numbers for its expression and not merely the integers. I may *count* the pounds to the nearest whole number; in order to *measure* the fat ever more precisely, I need those fractions.

However useful, the fractions retain under close inspection a certain unwholesomeness, even a kind of weirdness. For one thing, they appear from the first to be involved in a suspicious conceptual circle. An ordinary fraction is a division in prospect, with 1/2 representing 1 *divided* by 2. But the rational numbers were originally invoked in order to provide an account of division amidst the integers. The operation of division has been explained by recourse to the fractions and the fractions explained by recourse to the operation of division. This is not a circle calculated to inspire confidence. It is for this reason that mathematicians often talk of fractions as if they were *constructed* from the integers, a turn of phrase that suggests honest labor honestly undertaken. The construction proceeds in the simplest possible way. The fractions themselves are first eliminated in favor of pairs of integers taken in a particular order, with 2/3 vanishing in favor of (2, 3) and the somewhat top-heavy 25/2 in favor of (25, 2). The symbolic universe now shrinks—gone are those elegant fractions; and then it dramatically expands—pairs of integers come into existence. What is required to make this rather suspicious shuffle work is some evidence that the ordinary arithmetic operations by which fractions are added, subtracted, multiplied, and divided carry over to pairs of integers.

As, indeed, they do. Two fractions a/b and c/d are equal when $ad = bc$. The same number is represented by 2/3 as by 4/6 *because* 2×6 is just 4×3. High-school wisdom. But ditto for the pairs of numbers (a, b) and (c, d), whatever they may be. Ditto how? Ditto *by definition,* the mathematician simply saying that (a, b) is the same as (c, d) if $ad = bc$. And ditto again by definition when it comes to adding, multiplying, subtracting, and dividing pairs of integers, those pairs coming in the end to perform every useful function ever performed by fractions.

In this way, the rational numbers are emptied of one source of their weirdness—fractions; thus removed, those fractions are promptly reintroduced into the mathematical world on the reasonable grounds that if questions come up (*what are those damn things?*), they can always be answered (*pairs of integers*).

With fractions in place, the system of numbers in which they are embedded undergoes a qualitative change. The integers are discrete in the sense that between 1 and 2 there is absolutely nothing. There is not much more, needless to say, between 2 and 3. Going from one integer to another is like proceeding from rock to rock across an inky void. The fractions fill up the spaces in the void, with 3/2, for example, standing solidly between 1 and 2. There are now rocks between rocks—the void is vanishing—and rocks between rocks and rocks, with 1/3 standing between 1/4 and 1/2. The filling-in of fractions between fractions is a process that goes on forever. That void has vanished. The number system is now *dense,* and not discrete, infinite in either direction, as the positive and negative integers go on and on, and infinite between the integers as well.

In looking at the space between 1 and 2, swarming now with pullulating fractions, the mathematician, or the reader, may for a moment have the unexpected sensation of peering into some sinister sinkhole, some hidden source of creation.

chapter 3

The
Black
Blossoms
of
Geometry

GEOMETRY IS A WORLD WITHIN THE WORLD. THE INTEGERS AND THE fractions represent the numbers with which that world must be coordinated. But geometry is one thing, arithmetic another. Taken on their own, they remain alien, one to the other. Analytic geometry represents a program in which arithmetic comes vibrantly to life within geometry, and so describes a process in which an otherwise severe world is made to blossom.

Now, in its most abstract and consequently its most beautiful incarnation, Euclidean geometry arises out of nothing more than a collection of lines and points. Enter Mrs. Crabtree for a final, forlorn appearance. *You see,* she is saying, *a triangle is simply the interior of three mutually intersecting straight lines, and a circle is determined when a straight line sweeps around a point.* She pauses to survey the effect that this declaration has on the class. *Lines and points,* she says sadly. And then her features merge again into nothingness, leaving behind for only a moment an outline of her thin frame, an outline that tapers to a solitary point and disappears.

The program of analytic geometry is to evoke the numbers from the stubby soil of a geometrical landscape; it begins with a solitary line, something that lies in the imagination like a straight desert highway stretching from one blue horizon to the other. The traveler drifting down that highway, it is worth remembering, requires only *one* landmark to orient himself. Like the hero of innumerable westerns, he is heading *toward* Dodge City, or like the villain of those same westerns, *away* from Dodge City, Dodge City itself serving as the solitary point on the otherwise empty and lonesome stretch of road telling the cowpoke where he is going and the villain where he has been.

What is good enough for the cowboy is good enough for the mathematician. Looking at a given line, *he* picks a point to serve as a starting spot. That point functions as an *origin,* a source of things and a center of motion. *Hey you! Start here.* With an origin in place, picked out of the line by the mathematician's arbitrary but oddly compelling gesture, the mathematical line, like the desert highway divided dramatically by Dodge City, is itself divided into what lies to either side of the origin, the simple act of fixing an origin endowing the line with an eye-arresting structure where previously there was only something featureless as an egg.

Dodge City is, of course, a real place—the saloons, the brothel above the feed lot, the churches, ornate and mad in the evening sun; and the origin is a mathematical point, something that has sucked from the concept of a place its essential property, that of being *here* rather than there, the infinitely extended line itself balanced perfectly on that slim, solitary, and singular spike. But a point, it must be remembered, is *not* a number; holding place without size and arising whimsically whenever two straight lines are crossed, it is a geometrical object, a kind of fathomless atom out of which the line is ultimately created. Analytic geometry is a program to make the desert bloom; but if arithmetic is to be found here it can only be as the result of a deliberate assignment of numbers to points, a pairing of items that are incorrigibly distinct. The mathematician thus does not discover a number at the origin: He *invokes* one. Looking out over that linear landscape, the line bisected by a point, he assigns the number 0 to the origin, if only to convey the sense on the line already conveyed in the number system itself, that at 0 things have a beginning (0, 1, 2, 3, 4, . . .) and at 0 they have an end (. . . , −4, −3, −2, −1, 0).

One number has been made to flower and break black blossoms on the line; the rest of them may be made to follow and crack the stony soil.

In nature, some things are close (the lion and the tiger, cats both) and some things far apart (the tiger and the flatworm, different animals, different phyla even), the concept of *distance* one of the crucial, if generally hidden and obscure, instruments by which we assess the world and find our way within it. Distance is a concept with a thousand florid faces—there is emotional distance, intellectual distance, biological distance, psychological distance, geographical distance, moral distance, aesthetic distance, sociological distance—but in mathematics distance is defined by reference to a space of some sort and is thus a concept that requires, among other things, a fixed point, the question *how far?* prompting in turn the inevitable further question *from where?* On the line, at least, the *where* has already been specified. It is the origin and thus a place with a firm numerical identity at 0.

Given any *other* point on the line, *how far is it* now acquires a precise interrogative meaning, as in *how far is it from the origin to this very point*. With Dodge City or the origin burned into consciousness as a fixed point, *not far* is one answer to the question, and useful to the extent that it reminds us that distance is a qualitative as well as a quantitative concept; but in response to the additional question *how far is that*, arithmetic comes to the fore, if only to specify the distance in terms of minutes, miles, or meters, and so inescapably in terms of numbers.

Now among the numbers, 1 functions as a *unit,* an indestructible and bouncy atom into which every other number may be laboriously, but inevitably, decomposed. The number 10 is, after all, nothing more than ten of those 1's; and with the number 100,000, it is more of the same—1's strung out as far as the eye can see. This suggests that as numbers are multiples of some unit number, distances, too, are multiples of some unit distance, some fixed expanse functioning as the atom by which every other expanse is realized and then reckoned. And, indeed, distances on the open road or on the line *must* be multiples of some unit distance simply because every number is a multiple of some unit number and distances are measured in terms of numbers.

The monotonous miles, meters, and minutes of that open road may now be permitted decently to disappear, but distance remains as a concept and so, too, the concept of a unit distance. Having chosen an origin, the mathematician next chooses some fixed distance on the line to represent the unit distance, the process involving nothing more compelling than this character, the mathematician, holding apart thumb and forefinger and saying *that's about right*. The choice of a unit is arbitrary. The distance is *fixed* because it is a measure of distance from the origin. And it is a fixed *distance* because the mathematician is measuring spatial expanse. With a unit distance thus in place, a second number makes an appearance on the line. The point precisely one unit distant from the origin is assigned the number 1.

The line has now been made to blossom twice. The number 0 marks the point at which things begin; 1, the unit distance. No further effort is needed. That line blossoming in just two spots may now be seen to blossom everywhere, like one of those old-fashioned time-lapse movies in which a somnolent suburban garden, all drooping pansies and primroses, comes suddenly to vigorous and alarming life as the film is speeded up. Just as on the highway itself the distance between Dodge City and Wherever is expressed as a multiple of miles (*Dodge City? I reckon it's about ten miles, Marshal*), the distance between the origin on the line and any other point is expressed as a multiple of the unit distance. The number 2 blossoms on the line at the point two units from the origin, and 3 follows in turn. Every natural number is represented in just the same way. The fractions on this scheme play the role that they always play, 1/2, for example, denoting the point midway between 0 and 1. There are no surprises. Things are just as they seem. The scheme is simple.

If the positive integers and fractions indicate distance from the origin in one direction, the negative integers and fractions indicate distance from the origin in the other direction. It is here that the lucidity of a geometrical stage—its high desert light—may first be appreciated.

The negative numbers are, perhaps, the first of the great counterintuitive concepts of mathematics. A number representing quantity, it is troubling to think of *negative* quantities, things that are less than zero (although examples such as the novelist Bret Easton Ellis do come easily to mind). But on the line, the negativity of the negative numbers indicates nothing more than their *direction,* the fact that if the positive numbers are moving to the right, the negative numbers are moving to the left.

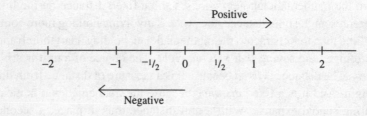

The number line

This elegant little exercise complete, the numbers have been inscribed on the geometric line, endowing the line with a living arithmetic content and being endowed by the line with a geometrical exoskeleton. Points on the line have now been assigned a numerical magnitude, and numbers a geometrical distance. It is possible to *measure* the distance between points and possible again to *see* the distance between numbers. Far from seeming strange, this interpretation of arithmetic and geometry strikes a deep, a resonant, chord of intuition suggesting that contrary to the historical development of these subjects, arithmetic and geometry are each aspects of a single, deeper discipline in which form and number are seamlessly matched and then merged.

chapter 4

Cartesian
Coordinates

WISHING TO TRAVEL THROUGH GOD-KNOWS-WHERE, I IMAGINE MYSELF
doing what I never do in real life: looking at a map. What I see on the
printed page are points indicating a variety of memorable places: There
is Plaatsville, home to the Plaatsville Gophers, there is the birthplace
of Asa H. Aberfawthy, inventor of the skinless frankfurter, and there
is the site of the world's largest processed-cheese factory. At the bottom of the map are letters going from **A** to **E**; at the side, numbers from
1 to 5. These are the map's coordinates.

This I know. Everything else is hopelessly unclear. I turn the map
aimlessly in my hands.

Where do you want to go, fellah?

I am now staring at the map in blind bafflement, suffused by a mounting fury, my wife smirking as the gas station attendant, all grease smells and a network of fine lines about his eyes, stabs at the map with his index finger.

Leper's Depot, we want to go to Leper's Depot.

From the huffy character to my right: *You want to go to Leper's Depot. I want to go home.*

With what I imagine to be a smile of superiority, the Master of the Pumps effortlessly reads the map's alphabetical index.

Lardvista, Lawrence, Lemis, Leper's Depot. Here it is. E5.

I count from 1 to 5 at the side of the map. I go from **A** to **E** along the bottom. I extend the imaginary lines. What do you know? There it is. Leper's Depot. Right where it is supposed to be, a demonstration, if any were needed, that any point on a map can be fixed by two co-ordinates and that every coordinate pair (letter and number) picks out a point on the map.

The gas station, with its flapping pennants and mummified attendant, serves to express the single luminous idea of analytic geometry. Map making and mathematics alike proceed by the identification of points or places with pairs of numbers. In the case of mathematics, the coordinates of choice are always numerical, if only for convenience, and created by the perpendicular intersection of two number lines. Such are the axes of a mathematical map. Their point of intersection at 0 represents the origin of the map, the system's rooted center.

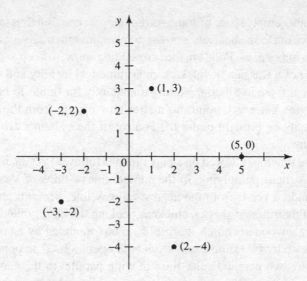

Cartesian coordinate system

A unified and simple system of measurement now comes into play. Distances from the origin along both number lines are marked in a common way and by common numbers—the ordinary integers and fractions. Above and to the right of the origin go the positive numbers; below and to its left, the negative numbers. These number lines, together with the space that they span, embody a Cartesian coordinate system, the system itself serving to depict the geometric plane. Gone are the towns and villages of a paper map; *this* map plays over an infinite panorama of points.[1]

Like an ordinary paper map, a Cartesian coordinate system is meant to provide information about the whole of the plane that it spans by means of its number lines. They may, those number lines, be imagined as a pair of perpendicular railway tracks plunging through the vast

[1] The assumption that space contains infinitely many points is so commonly and so easily made, that it is tempting to forget that it is in frank conflict with the atomic hypothesis of modern physics, the idea that the decomposition of space proceeds only for a finite number of steps and ends with the display of only a finite number of fundamental or elementary particles.

and somber empty space of some surrounding steppe. But thus far numbers have made an appearance on the number lines themselves and only on the number lines. Points in the surrounding snow, to keep to the new imagery of a Russian novel, lack an arithmetical identity and so simply sit out there in Siberia, waiting impatiently for things to hurry up and begin. Yet every point, no matter how distant from the origin, may easily be brought under the control of the system's arithmetic apparatus.

The scheme followed on a paper map is followed yet again on the plane. A rogue point lying off the number lines—think of Vasilyevo, five hundred versts from the nearest railway line, peasants sitting in front of their ruined shacks, chickens pecking the hard ground, in the distance a wooden church shaped like a box mounted by an onion— is assessed by two simple and sequential operations. The point is bisected by two perpendicular lines moving parallel to the coordinate axes themselves. Where they, those parallel lines, intersect the number line, the intersected numbers are assigned to the point.

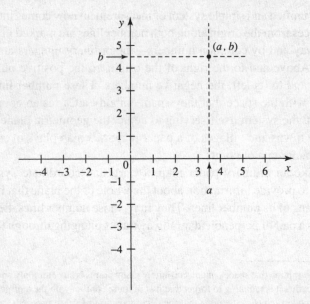

Assigning coordinates to
point (*a*, *b*) in the plane

This is identification by proxy, Vasilyevo gaining a lively sense of itself from its relationship to those impossibly distant railway lines that might take a traveler to Moscow or St. Petersburg. Simple as the scheme is, it works for every point; with number lines in place, the geometric plane acquires a healthy glow, as each of its infinitely many points is endowed with a unique address—its *coordinate*—and thus with a unique identification.

An origin is in place; *there* the coordinate axes intersect. The number lines have acquired a doubled identity, with distances along the line assigned numbers, and numbers now standing in to mark distances. The plane as a whole has attained a precise arithmetical character, its points associated with pairs of numbers. The poignant particularity of a world in which every place is different from every other place in innumerably many ways—this has been lost. In a Cartesian coordinate system, points are all of them alike except for their addresses. But a coordinate system is sufficient to express the concept of distance on the line and beyond that of distance in the plane. The metaphysical experience of rootedness has been given a mathematical echo in the concept of an origin, and the very essence of what it means to be one place rather than another has been expressed by the system's points and the numbers that identify them.

French philosophers, take note.

Compelled by a Form of Words

Students who need not be persuaded that gender studies are good for something often ask innocently whether analytic geometry is good for anything. *But, of course* is the short answer. The long one, too. Analytic geometry allows the mathematician to describe geometrical figures by equations and so begin the work of bringing forms and shapes and soft sensuous curves under the control of symbols. Consider a straight line, for example, one suspended in midair and dutifully propagating itself throughout all of space. What can one say of this line except something on the order of *there she goes*? But imagine it, that line, passing through a Cartesian coordinate system and ascending upward for parts unknown.

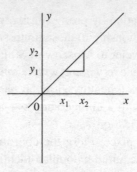

Straight line passing through the origin

It is plain from the picture—it may be taken as an assumption—that for every unit the line ascends, it moves one unit to the right. Moving up, it goes from y_1 to y_2; and moving out, from x_1 to x_2. (The subscripts serve as general placemarkers, x_1 designating the first stop along the x-axis, and hence the first place, and x_2 the second stop and so the second place.) The ratio of distances covered thus in two directions, vertical to horizontal, is the line's *slope:* $(y_2 - y_1)/(x_2 - x_1)$, to put the matter into symbols, where $(y_2 - y_1)/(x_2 - x_1)$ is in this case equal to 1. It is plain again from the picture that this straight line crosses the y-axis at the origin. The line's slope and crossing point are fixed. And these are simply the descriptive circumstances attending this line, its circumstantial identity.

Back in the rosy-fingered dawn of delight, analytic geometers discovered that a straight line in a Cartesian coordinate system may be described, and described *completely,* by an equation of the form $y = mx + b$. The six symbols have an expected, if imperfectly remembered, meaning. Late-alphabetic variables such as x and y act as ordinary English pronouns, bits of grammar indicating where something is unknown, even as *He did it* leaves in the dark both *who* he is, and *what* he did. Solving an algebraic equation (or any equation at all) is a matter of determining who *he* is, or what he *did*, this on the basis of various clues left lying about the equation. The buoyantly bursting b, on the other hand, functions as a proper name and represents the particular point at which a line intersects the y-axis; m is another proper name, this one denoting the line's slope. Variables in an equation are variable and take on different values in the course of the same equation, but the proper names are fixed and frozen.

Every point on a line domesticated by a Cartesian coordinate system corresponds to two numbers. These are its coordinates, the mark of its domesticity, the one corresponding to distance along the x-axis, the other, to distance along the y-axis. The equation $y = mx + b$ says that given a value of x, *it* can determine unequivocally a value of y. This is a remarkable thing for an equation to say; it is possible for the equation to say it only because the system of constraints that it expresses is screwed tightly enough to specify the unknown values of y.

Suppose thus that $x = 423$. This vagrant bit of information serves only to direct your attention to the cold of interstellar space, where a point on the line occupies a position 423 units from the origin along the x-axis. The information supplied supplies half the point's address. The y-coordinate is yet missing. And *this* the equation returns: $y = 423$ and y *must* be 423, since $m = 1$, $b = 0$, $x = 423$, and y is identical to $mx + b$. Ordinary English rarely offers a sentence so tightly wound as to make the identification of an unknown inescapable. The *he* in *He did it* may refer to anyone from Charlemagne to Harry Houdini, and only the context of utterance and a refinement of the description itself serve to narrow the field, *He did it his way,* for example, just possibly specifying a toupeed Frank Sinatra.

What $y = mx + b$ does for one point, it does for them all and so draws the curtain of its command over infinitely many coordinates. *Infinitely* many! A straight line may be more than the simple sum of its points, but whatever its identity, the line is *expressed* by the points that compose it. Suspended in space and plunging bleakly toward the double darkness of distance in two directions, it is the line itself that now falls under the control of a formula, a finite form of words.

In the fairy tale, Aladdin stood before a secret cave and sought to murmur an enchanted word. Poor schnook. He wished only for booty. With *words,* the mathematician controls the illimitable.

Chez Descartes

The restaurant is always open, copper pots gleaming against the dark smoked wood walls. The mathematicians seating themselves comfortably after greeting the owner's wife, *they* know René Descartes as a mathematician: his role as a philosopher they pass over with an embarrassed silence. Even if it provokes only an indulgent, an ironical, murmur of appreciation, it is well to remind them that Descartes is the first of the modern philosophers, and the greatest *because* he is the first.

Descartes introduced the method of doubt into philosophy and so made philosophers begin their work with skepticism—the problem posed by his blunt, penetrating question: *how do we know?* How do we know *what? Anything at all.* His *Meditations* remains the text from which modern philosophers never quite escape, however much they may imagine that they have liberated themselves from its deep hypnotic influence. Drawing a distinction between the mind and the body, and thus between the world of experience and the physical world, Descartes argued prophetically that it is the mind and *not* the body that is intuitively known, easily accessible. The success of the sciences seems to suggest the contrary, but it is a success achieved at great intellectual cost and over a very long time, and in the end we may all of us find ourselves trooping back to Chez Descartes, to sit with the mathematicians in the dark and drowsy room in order to contemplate the mind contemplating itself as the smell of goose fat fills the air.

Descartes was born in 1596 near Tours, where French is said to be spoken with exceptional purity and where, I am reminded by one of the mathematicians, they know how to eat well. He was delicate as a youth, and indulged by his father, and passionately preoccupied by mathematics. He remained a valetudinarian throughout his life, a man capable of taking to his bed at the sign of a pimple; yet in his youth his biographers have him tramping off to war as a common soldier, one idiotic campaign after another. He fought in the Battle of Prague; he signed up for military duty with the Elector of Bavaria. He seemed genuinely to like the soldier's cold damp tents, water dripping down the canvas walls, the crude cannon's flash, the roar and clash of combat as tense violent men took to the muddy fields.

And through it all, he carried on his passionate intellectual affairs, able somehow to reconcile an interest in the ineffable with the brutal squalor of a soldier's life. The chief, indeed, the only, *chef* on one historical stage, that of analytic philosophy, Descartes is also a leading character on another far busier intellectual stage, that of mathematics, where he played successfully the role of a great and visionary mathematician, his masterpiece *La Géométrie* suggesting a dramatic and far-reaching synthesis forged between algebra and geometry. He came to mathematical maturity in the early years of the seventeenth century, and so missed the moment of creation available to Leibnitz and to Newton; but with respect to the calculus it was Descartes who was vouchsafed the vision of an ancient geometrical landscape infused with arithmetical life, and so it was Descartes who furnished the design without which they could not have completed the arch.

In his death as in his life, Descartes was unique: he was the only great thinker to die of discomfort. Summoned to Stockholm as a royal tutor by Queen Christina, he discovered to his horror that the queen, a vigorous young amazon, expected lessons in geometry early in the morning. The prospect was dismal, the hour ungodly, the city steeped in snow, with a bitter Arctic wind blowing over the dark sullen waters—Stockholm is virtually an island. Descartes despaired of his duties. He became ill with a pleural inflammation, pneumonia, no doubt, and died in his fifty-fourth year, coughing out his lungs as well as his life six years after Newton was born.

chapter 5

The Unbearable Smoothness of Motion

AND YET THERE IS ALWAYS A YET.

The transposition of a living body in space has its mathematical echo in the purely imaginary progression of successive distances along a number line. Going down that desert highway I am increasing my distance from an origin, the mathematical line serving to mirror my ecstatic forward rush. It is in that ecstatic forward rush that experience reveals one of its curious and haunting aspects. Whatever the journey, I may interrupt it simply by stopping; I may surge or shuffle; but so long as I am transposing myself from place to place, the experience of motion is *continuous* in a sense of the word that is pregnant without yet being well defined.

The unbearable smoothness of motion.

The geometrical line reflects the unbearable smoothness of motion perfectly; between points, there are points, those points falling in on themselves so that the line as a whole forms a *continuum,* an ancient mystic image of things at the margins of distinctness, a perfect expression of the passage we make from one place to another, or from one time to another, the experience of continuity suggesting that at some level there is only *seamlessness,* an intimation of the greater unity in which the immemorial distinction between subject and object disappears and the soul swims out to join the great ocean of being.

Yet the numbers are pretty hard-edged characters; each possesses a defiant sense of its own individuality, and none of them seems inclined to do much swimming toward the ocean of being. Or anything else. If points on the line find their separate identities a burden, the numbers positively *revel* in their individuality. This circumstance may provoke a squeak of suspicion, a sinister hunch that the line and the numbers inscribed upon it are in some way discordant. And although these remarks are delivered by a shrug of intuition, the shrug is backed up by an ancient argument.

The *Ifs* Accumulate

A theorem attributed to Pythagoras affirms that if a and b are the sides of a right triangle and h its hypotenuse, then $a^2 + b^2 = h^2$. The theorem embodies a striking *fact* about right triangles: whatever their particular configuration, *this* simple numerical relationship will hold among their sides. If $a = 3$ and $b = 4$, $a^2 + b^2 = 25$, and h must therefore be 5.

And so it is, the Pythagorean theorem embedding the waywardness of the world in an incorruptible set of conceptual constraints.

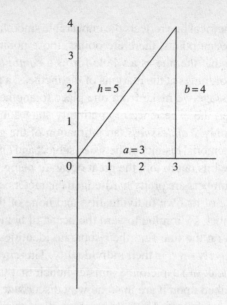

The Pythagorean theorem

But suppose now that *a* and *b* are 1. The triangle answering to the supposition appears unremarkable. Its legs are each one unit in length. The thing seems somewhat squat. But what of *h* amid all this ordinariness? Among other things, *h* expresses the extent of a fixed and hopelessly prosaic distance in the real world. And if *h* is a distance in the real world, it is also a distance on the number line, a fact that may be seen by rotating the triangle so that its hypotenuse coincides with the axis of the number line itself. Thus inscribed on the number line, the endpoint of the hypotenuse is at precisely *h* from the origin.

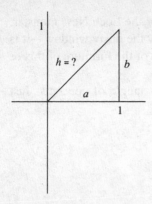

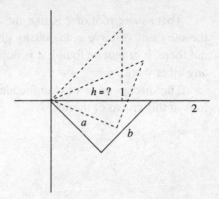

THE PYTHAGOREAN THEOREM
WHEN *a* AND *b* ARE 1

THE HYPOTENUSE *h* AS AN EXTENT
ON THE NUMBER LINE

So? What then *is* *h*? A distance of what *magnitude*? The question belongs to the great good-hearted congregation of questions asked by the hopelessly lost.

Leper's Depot. Is it some distance from here?

Ayuh, some distance.

Can you tell me how far?

It would be intellectually repugnant to learn that although Leper's Depot *is* some distance from here, it is a distance that *cannot* be correlated with any number.

Can't say, bub. No telling.

This laconic response, which evokes one of those absurdist dramas so popular in the fifties, is nonetheless *appropriate* in the case of *h*, the suspicions and surmises now collecting themselves into a flat and sullen statement: *there is no way of telling*.

The overall argument is very simple, very compact, and very powerful. The Pythagorean theorem says that $a^2 + b^2 = h^2$, and it says so for *any* right triangle. If *a* and *b* are 1 and thus $a^2 + b^2 = 2$, *h* is then the number that when squared (or multiplied by itself) is 2. These trim and tidy inferential steps suffice to take the reader to the very edge of doom. If $a^2 + b^2 = 2$, h^2 must be 2 and *h* itself $\sqrt{2}$.

But no such number exists.

That square root of 2 is like the Yeti or the Loch Ness monster, the snows of yesteryear, the dusky ghost by the dusty window—it is not there, it cannot be found, it is not a part of the furniture of this or any other world.

The discussion is now embedded in a tangle of concepts. Just look at this crown of thorns.

chapter 6

Y_0

THE SQUARE ROOT OF 2? IT DOESN'T EXIST? *You're putting me on, right?*—this said with the tone of incredulity with which on ordinary occasions we treat an old friend's announcement that he is about to depart for an ashram. The thing is puzzling. It puzzled the Greeks, and it puzzled mathematicians who came after the Greeks. It puzzled mathematicians filing down the centuries, God-intoxicated Hindu sages writing in the shadows of gorgeous temples, bearded Arabic scholars fingering their caftans, profit-eyed men of the Italian renaissance.

But there it is. The ancient proof is unassailable and proceeds with the irritating authority of a waking nightmare, one of those squalid episodes, say, in which a disheveled taxi driver, unshaven and nonecked, turns and addresses you insolently, but with perfect and disturbing confidence:

"*Yo,* man. Suppose that $\sqrt{2}$ does exist. I mean like do yourself a favor and just suppose it. Then $\sqrt{2} = p/q$, where p and q are these here numbers. Am I right?

"But p and q, they can't both be *even* numbers on account of if they were you could always divide through by 2. I mean if they was like 2/4 or something, you could always say that's the same as 1/2 and use that instead. Am I right?

"*Yo,* now $\sqrt{2} \times \sqrt{2}$ has got to be the same thing as $p/q \times p/q$. I mean, fellah, that's just what the square root of anything is.

"So does $\sqrt{2} \times \sqrt{2} = 2$ or what? And $p/q \times p/q$, hey, that's just p^2/q^2.

"So what you got, you got $2 = p^2/q^2$, and then you got $2q^2 = p^2$ on account of you multiply both sides of the equation by q^2. You can do that and all.

"So that means p^2, it's got to be even—am I right?—on account of you divide both sides of the equation by 2.

"p^2 being even, it means that p itself has got to be even. Am I right? That's on account of odd times odd only gives you odd.

"*Hey,* come to think of it, that's true in the taxi business, too.

"So follow me on this. We take $p/2 \times p/2$, we get $p^2/4$.

"*Yo,* I *am* coming to the point, and besides with this traffic we ain't going anywhere. Look back at what I done. We know p^2 is just $2q^2$. So if p^2 is divisible by 4, same thing for $2q^2$, them being equal and all.

"You with me, or what? I'm asking is on account of it's all over. $2q^2$, it's divisible by 4, you got to figure that q^2 is divisible by 2. Am I right?

"Which means q is divisible by 2.

"Which means p and q are *both* even. Now the reason you're sitting there nodding your head like a dummy is that five minutes ago I tell you that p and q can't both be even.

"So what's it going to be, Chief?"

What indeed?

Poor Fat Things

The argument just given proceeds by remorselessly flogging an assumption toward a contradiction—*reductio ad absurdum* as it is known
cheerfully in the trade, especially by those doing the flogging. This particular *absurdum* evokes something like a feeling of fretfulness. Sixteen has a square root in 4, and 1/4 a square root in 1/2, but 2 has no
square root whatsoever among the rational numbers, although it would
appear that 2.25 has a square root in 1.5. *Fretfulness?* That is not quite
the right word. An ancient impediment to understanding has come
shambling out of the historical mists, dragging green slime behind it
and snorting wetly. *Impediment?* Not quite the right word either. The
taxi driver's argument may be given again and again. There are plenty
of square roots beyond the square root of 2 that cannot be expressed
in terms of the rational numbers—the square root of 3, for example.
Like plush that under strong light reveals a series of alarming moth
holes, the familiar number system is filled with strange gaps, places
of reverberating emptiness. And the word for *that* is weird.

The square root of 2 forced the Greeks to the contemplation of *incommensurable magnitudes*—distances on the line that could not be
correlated with any number. These are unlovely objects, those numberless distances, if only because like hairless dogs they exhibit their
deficiencies so defiantly. The discovery of incommensurable magnitudes provoked a crisis among Greek mathematicians committed (as
most mathematicians are) to the supremacy of numbers. The story is
told of a mathematician rash enough to make the argument in favor of
numberless distances on board ship. The scene is irresistible: the brilliant Mediterranean sky, the wind in the starched sails, the heavily mus-
cled men for the moment pausing at their oars, elbows on their knees,
as a short, squat stranger firmly declares that contrary to every expectation some distances cannot be measured by means of the natural or
rational numbers. The oarsmen look to one another and then, it is satisfying to recount, rise unbidden, seize this presumptuous dweeb by
the scruff of his tunic, and, even as he pompously protests that he is
after all a mathematician, pitch him into the wine-dark sea. The crisis
they provoked, the Greeks never resolved. In Eudoxus and in Euclid,

incommensurable magnitudes make an appearance as incommensurable magnitudes, strange numberless objects. Ratios of such objects are taken and a scheme of geometry created, but in the end, there the poor fat things sit: obscure, implausible, and bizarre.

The great Hindu and Arabic mathematicians of the Middle Ages took quite another tack. Whatever incommensurable magnitudes might be, they treated such things *as if* they were really numbers—*irrational* numbers, the *irrational* a nice inadvertent touch signifying the madness loitering about the very notion—and learned many tricks by which such numbers might be manipulated. In the twelfth century, for example, Bhāskara demonstrated correctly that $\sqrt{3} + \sqrt{12} = 3\sqrt{3}$, an achievement, I might add, utterly beyond the collective intellectual power, say, of the English department at Duke University. (It is pleasant to imagine members of the department sitting together in a long lecture hall, Marxists to one side, deconstructionists to the other, abusing one another roundly as they grapple with the problem.) But neither Bhāskara nor anyone else ever made clear what items such as $\sqrt{3}$ *were*. The symbols resisted, as symbols so often do, any attempt to invest them with meaning. Sitting in their perfumed gardens, those thousand and one Arabian mathematicians carried out their calculations with a charming and insouciant assurance that all that gibberish actually made sense.

Not that anyone else did any better, the high medieval gibberish of Arabic mathematics appearing in Italy, France, and England as an inexpungably vital but irremediably vulgar weed. And the curious counterintuitive thing is that it didn't matter. The commonplace view of mathematics as a discipline *consecrated* to the ideal of precision has very little to do with mathematics as it is lived. Between 1500 and 1800, the great central stage of European thought is crowded with babbling and arguing figures—Cardano, Stifel, Pascal, Descartes, Wallis, Barrows, even Leibnitz and the sainted Newton—saying one thing but writing quite another, agreeing in solemn convocation that irrational numbers are a fiction (almost a certain sign of bad faith, in mathematics or anything else), and then applying that fiction to numerical problems and like Bhāskara miraculously getting the answer right, the work involved in the creation of the calculus a matter evidently capable of being conducted *without* being clarified, the mathematicians following the trail of some central truth so suggestive as to make delicacy of distinction a pointless distraction. *Allez en avant et la foi vous viendra,* d'Alembert was reputed to have said: *Go and advance and faith will come to you,* advice that with respect to the calculus is still offered piously to undergraduates (with, I might add, entirely predictable results).

The story could, of course, be continued into the byways of a thousand biographies. Let me convey an impression instead of what history has already conveyed: a forward rush of mathematical progress informed by a haunting and retrogressive sense that two ancient objects in human experience, the straight line and the numbers themselves, are somehow hopelessly discordant, the sense of dislocation all the more pressing and all the more poignant in virtue of the conviction, one shared by almost every mathematician, that the line should express the numbers and the numbers should represent the line, and that both expression and representation should be perfect and complete.

chapter 7

Thirteen
Ways of
Looking
at a Line

When the blackbird flew out of sight
It marked the edge
Of one of many circles.

WALLACE STEVENS

A SIMPLE STRAIGHT LINE, TENDING TOWARD THE INFINITE IN TWO DIREC-
tions, is an ancient object in human experience, the first of the great
abstractions from sensual life.

One way of looking at a line.

The edges and surfaces of the natural world tend irresolutely toward the bloblike; immersed in birth and decay, the biological world
is filled with warm and annealing surfaces, swelling up, curved, amorphous, the whole of creation organized but chaotic. A straight line has
the purity of any object that does not deviate. There the thing hangs,
severe as a sword blade.

Two ways of looking at a line.

And?

The line is *infinite,* composed of infinitely many points.
And?
The line is *dense.* Between any two points on a line there is a third.
And?
The line is *ordered.* Points on the line do not simply sit in an indiscriminate jumble. Any given point comes after the one before and before the one after.

And?

The line is *continuous.* There is no point at which the interior structure of the line fails to cohere, revealing an abysmal emptiness, the backside of the beyond.

Six ways of looking at a line.

And now an obvious but difficult question: *Which* of these ways of looking endows the line with its mysterious distinctiveness as an object of thought; in what does the *essence* of the line reside? Philosophers may now have the stage for thirty seconds:

Its coherence. Yes, absolutely. Its coherence.

The way it sort of hangs together and all.

The fact that like whenever you have two points, there's a point in between them?

It's distinctive on account of its continuity. If you were a bug on the line, you wouldn't fall in between two points or something. This is very hard to explain.

In one respect, the philosophers have pointed the way toward a detour that need not be taken. Mathematicians from the time of the Greeks had sought to explain the essence of the line by means of the density of its points. Physicists too, Ernst Mach in particular, arguing that the line is continuous just because between any two points there exists a third. They were wrong, those mathematicians, and so, too, was Mach. Between any two rational numbers (any two fractions), there is another rational number (another fraction). The proof? Take the sum of those fractions and divide by two. This, their arithmetic mean, will fall neatly between the two original fractions. The rational numbers, *like the points on the line,* unfold to reveal ever more of themselves in the spaces between themselves. And yet the line is richer than the rational numbers. It is simply not richer in *this* respect.

Six ways of looking at a line and six ways have been used up.

Traveling through space, it is my body straddling a Harley that divides the desert highway into What Has Passed and What Is to Come, thus creating a shifting sense of spatial *position*. A straight line is like a blue highway, and a mathematical point is like a geographical place, something separating what has come before (the dry desert, the smell of sage in the spring air, the countryside so strange somehow) from what lies ahead (the mountains, waterfalls on their flanks, spring flowers on limestone meadows).

Seven ways of looking at a line.

Traveling through time, it is my consciousness that divides my life into What Has Been and What Is to Be, the events that have taken place receding like pale repetitive scraps, the events that are to come lined up stolidly in the future, waiting their turn to take a shot, my consciousness of time passing creating a shifting sense of temporal *position,* an impression of a particular time. A straight line is like a lived-in life, and a point is like the present moment, a local eruption serving to separate what has come before (childhood, youth, maturity) from what lies ahead (middle age, old age, the abyss).

Eight ways of looking at a line.

The present time, and so a sense of *now,* divides the temporal continuum; the present place, and so a sense of *here,* the spatial continuum. Time and space alike are *severable,* the severing dividing time into Before and After and space (thinking now of a highway) into Back There and Up Ahead. A line is like a highway and like a lived-in life.

Nine and ten ways of looking at a line.

Severability is a subtle feature of experience, the color between two colors. In his remarkable essay *Continuity and Irrational Numbers,* the nineteenth-century German mathematician Richard Dedekind wrote with a sense of dawning discovery that it was severability (this alone) that gave the line its essence, its power. Let us suppose, Dedekind supposed, that at a point the geometrical line is in imagination cut. The result of the cut just made is a division of the line into two segments, **A** and **B**. Every point in **A** is to the left of any point in **B**.

Dedekind cut of the line at point *P*

The metaphor of a cut has meaning, of course, because the line is ordered by the placement of its points (as a highway is ordered by the placement of its cities and towns); but the English word "cut" fails to suggest the meaty decisiveness of the German *geschnitten,* the energetic suggestion of action undertaken, as if the line were actually being snipped by a pair of heavy shears.

Every point on the line determines one and only one cut, one place where the line may be separated into two wings, and this is a feature of the line obvious enough to be overlooked.

Eleven ways of looking at a line.

But every cut on of the line is made at one and only one point, one spot where the wings of the line coincide, and this, too, is a feature of the line subtle enough to be overlooked.

Twelve ways of looking at a line.

In the sternest of mathematical objects, there are aspects of life itself—the sense of the present, the sense of place, and the ways in which time and space are cut. Life is like a line and a line is like life, the metaphor fading in favor of what it means and what it means fading in favor of the metaphor.

Thirteen ways of looking at a line.

chapter 8

The Doctor of Discovery

IT IS BEST TO THINK OF DEDEKIND AS A GREAT DIAGNOSTICIAN, A DOC-
tor of discovery. The facts are in order; but the *facts* have always been
in order. The *facts* have been in plain sight for more than two thousand years. Here they are, those facts. Some distances on the line cannot be correlated with any natural or rational number. And the numbers
contain gaps, places where there should be something but where there
is nothing instead. Crude ear trumpets in hand, other doctors have already drawn the obvious conclusion: the line is inherently a *richer* object than the numbers. There is something about the line, one of them
says, some kind of continuity, some special property, some thing or
aspect, some feature or condition; but when it comes to specifying what
that thing, aspect, feature, or condition is, he lapses into silence.

Dedekind listens quietly, his thin lips pursed. Then he says: "The comparison of the rational numbers with a straight line has led to the recognition of the existence of gaps, of a certain incompleteness or discontinuity of the former, while we ascribe to the straight line completeness, absence of gaps, or continuity."

His tone of voice is professorial; he speaks with grave courtesy.

"In what then does this continuity exist?" he asks rhetorically. The other doctors look alert.

"Everything," Dedekind says, his voice still calm, still serious, surprisingly musical, "must depend on the answer to this question."

Some of the doctors nod their heads.

"For a long time, I pondered over this in vain, but finally I found what I was seeking."

Yes?

Dedekind turns from the circle of attentive faces and palpates the bedridden patient with his elegant and sensitive fingers. Withdrawing his hands with a murmur, he points to the profoundly tender spot he has just touched and the mysteriously twitching incondite nerve behind it.

"It consists of the following," Dedekind says. "We know that every point p of the straight line produces a separation of the line into two portions such that every point of one portion lies to the left of every point of the other portion." Dedekind holds his hands in the air, the palms rotated outward, to suggest the line broken into two parts. His hands resemble bird wings.

But my dear Doctor . . .

"I find," Dedekind continues, "the essence of continuity in the converse."

The converse?

"In the following principle"—and here Dedekind's voice becomes more clearly focused, compelling. "If all points of the straight line fall into two classes such that every point in the first class lies to the left of every point in the second class, then"—there is a pause for emphasis, Dedekind raises his finger—"there exists one and only one point which produces this division of all points into two classes, this severing of the straight line into two portions."

The tense doctors look to one another again.

"By this commonplace remark the secret of continuity is to be revealed."

But this is so obvious.

"Yes," Dedekind says. "Most people may find its substance very commonplace." He shrugs gently. "I am glad if everyone finds the principle so obvious and so in harmony with his own ideas of a line, for I am utterly unable to adduce any proof of its correctness."

A sibilant murmur arises in the room.

"Nor has anyone else this power," Dedekind quickly adds. "The assumption of this property of the line is nothing else than an axiom by which we attribute to the line its continuity, by which we find continuity in the line."

But, Doctor, are you saying that continuity is something we create? Surely it is something we discover?

Dedekind shrugs as if to indicate that the difference does not matter much.

And the rational numbers? What of the rational numbers?

"Every rational number r," Dedekind says, "effects a separation of the system into two classes A and B such that every number of the first class is less than every number of the second class."

And what role does r play, this number?

"It is either the largest number in A or the smallest number in B."

Yes, surely that is true, someone says.

"And in this respect," Dedekind adds, "the rational numbers are like the line."

The doctors look at Dedekind expectantly.

"But it is easy to show," he goes on to add, "that there exist infinitely many cuts not produced by a rational number."

Not produced by a rational number?

Dedekind nods his head, the sandy hair sleeked back. "Let us consider the number 2, for example." He removes his spectacles and for a moment massages the bridge of his nose between his tense fingertips. "If we assign to one class B every positive rational number whose square is greater than 2, yes? and to another class A all the other rational numbers, this separation forms a cut. After all, every number in A is less than any number in B."

The doctors lean their heads forward; they appear to agree.

"But this cut," Dedekind says slowly, "is produced by no rational number."

This I do not understand.

"It is easy to show," Dedekind says, "that there is neither in the class *B* a least, nor in the class *A* a greatest, number."[1]

For a moment the room is quiet.

"In this property," he adds, "that not all cuts are produced by rational numbers consists the incompleteness or discontinuity of the domain of all rational numbers."

The room, a smell of disinfectant and carbolic acid in the air, is absolutely still.

Well, yes, one of the younger men finally says.

Complete at Last

The line is in some sense richer than the numbers that are used to represent it, and this is an old, an inconvenient, fact; but Dedekind's diagnosis goes beyond a revisiting of such facts in order to display the long-hidden source of the discrepancy between line and number. Every rational number produces a cut among the numbers; but some cuts answer to no rational number and in this respect—*this alone, no other*—the numbers and the line are different. Dedekind's calm but profound investigation succeeds as an act of intellectual liberation because it connects a particular fact—that some distances cannot be measured by any rational number—with the much larger, the more general, fact that some *cuts* cannot be made at any rational number.

It is the strength of Dedekind's diagnosis that it suggests its own remedy. If the rational numbers are filled with gaps, new numbers, Dedekind urged, are needed to make good the deficiencies. Arabic and Hindu mathematicians may now be heard rumbling from behind the Beyond, where they are tapping their ivory-headed walking sticks in indignation; they are suggesting that *whatever* this Dedekind may suggest, *they* knew it all along. But mathematicians before Dedekind had

---

[1] Easy, that is, if you know what you are doing. A proof is given in the appendix. Dedekind, of course, knew what he was doing.

simply invoked the irrational numbers with a certain hearty careless-ness, trusting in their superb intuition to get things right. In Dedekind's diagnosis, new numbers arise as the result of an *informed* act of creation.

To speak of numbers in creation is to revisit a familiar conceptual stage. The natural numbers 1, 2, 3, 4, . . . , come unbidden as the sta-ples of a common human culture. Everything else in some way involves an act of fabrication. The example of the fractions—that odd business with pairs of numbers—suggests that like the sleek origami swan that emerges surprisingly from folded paper, new numbers can be made from those that are old. But the irrational numbers are strange con-ceptual objects, and as one might expect there is no *simple* way in which they can be defined. On the other hand, new numbers may be brought into existence wholesale, by means of a single generous ges-ture. I am temperamentally with those who favor wholesale methods (perhaps owing to a *shtetl* ancestor specializing in dry goods), the large dark nothingness on the line exorcised all at once by a single shaft of light.

The axiom that achieves these aims is surpassingly spare. "When-ever, then, we have to do with a cut *A* and *B*," Dedekind writes, "pro-duced by no rational number, we *create* a new, an *irrational* number." These may seem desultory words, but Dedekind is able to paint the por-trait of this new number precisely and so at least to supply the linea-ments of the desired miracle. It is to be a number in *A* greater than any other number in *A*; and thus a number less than any number in *B*. The axiom itself serves to *compel* such a number into existence. Given any cut of the numbers into two camps *A* and *B*, *there exists,* the axiom says—there *must* exist, the mathematician adds—one and only one number in *A* larger than any other number in *A*, the imperious *there exists* bringing something new into the world and so allowing the mathematician to share in the general mystery of creation. In the case of rational cuts, the axiom ratifies what is evident: the rational cuts are made at the numbers. But where before there was nothing more than an emptiness answering to the square root of 2, a new number now appears, a Dark Prince, an object utterly unlike any rational number, one flushed from the shadows and full of brooding mystery.

Dedekind published the results of his research in 1872 and so within the memory of the very oldest widow of a Civil War veteran, and I mention this in order to connect by some living tissue this moment with that one. The calculus had already been in existence for more than two centuries in 1872. If the calculus is much like a cathedral, its construction the work of centuries, it remained until the nineteenth century a cathedral suspiciously suspended in midair, the thing simply hanging there, with no one absolutely convinced that one day that gorgeous and elaborate structure would not come crashing down and fracture in a thousand pieces. Dedekind's axiom is *logically* among the fundamental affirmations of the calculus. With the axiom in place, the cathedral has a foundation. An assumption has been evoked to dispel a mystery.

Beyond the natural numbers, then, beyond the integers, and beyond the fractions, lies a still further class of numbers, great, brooding, and mute. These are the irrational numbers that long ago Arabic and Hindu mathematicians played with in their perfumed gardens. The system composed of the natural numbers, the integers, the fractions, and the irrational numbers acquires a new identity as the *real number system,* the *system* a seriously meant sign of its scope, one recalling Newton's *system of the world,* and the *real* a sign that *these* numbers are the drawn iron screws by which the world's scaffold is erected.

A Gabble of Ghosts

Speaking as I have been of Richard Dedekind and central Europe, let me at least set the stage. The apartment building is massive and made of stone. An unobtrusive bronze plaque indicates that its foundations were laid in the thirteenth century. A wide circular stairway, the steps worn smooth, leads from the courtyard to the upper floors. The apartments are themselves large, filled with large rooms. The furniture is heavy, made to last for generations, the sofa plump with red down cushions, the easy chair—an antimacassar on the arm and a deep indentation in the plush where someone has pressed his head—mounted on ornate bear-claw legs, an enormous pine armoire along the far wall, the area rug a muted rufous and gold Persian, the upright chairs by the table constructed of dark stained wood, the long dining room table heavy and redolent of oil, gas jets on the wall, a series of solemn family portraits in brown and white, an open piano in the parlor, a six-foot Blüthner, the legs carved and voluptuous. The kitchen is far away, on the other side of the house, home to giggling maids. The bathroom is tiny but immaculate, a closed closet consisting of a single mournful toilet. Outside, the heavy central European air is humid, and the sky is low and eternally gray.

Richard Dedekind was born in 1831; it is *his* head pressing against the plush of that central European easy chair and his long elegant fingers massaging the skin above his temples, a leather-bound book folded spine-up over his knee. We are in Germany, far from the sea, and Dedekind is drawn from the central European atmosphere: he is a man of the middle distance. Although his talent was evident from an early age, Dedekind passed his professional life as an instructor at a technical high school in Brunswick, the place of his birth. Nineteenth-century German *Hochschule* were rather more demanding institutions than contemporary American high schools, and Weierstrass, too, taught for some years in similar circumstances before he was awarded a university position. Still, high schools are all the same, superficial variants of some central high school which is located in Hell. Dedekind's professional exile has prompted his biographer's eyebrows to hoist themselves upward. It is hard to think of another mathematician of his stature who failed so conspicuously ever to obtain a university position. And yet judging from his correspondence, Dedekind was on good terms with the great mathematicians of his day. Reserved but unfailingly courteous, he may be seen behind his own handwriting, a man of calm judgment, his intelligence open, lucid, and affirmative. He never married, living with his sister. Such family arrangements appear to have been more common a century ago than they are today. His habits were said to be regular.

When I was young, I learned the calculus from an inspired little book entitled *Quick Calculus*. Like so many such books, it took pains to assure me that appearances to the contrary the ancient discipline I was endeavoring to master was actually pretty easy. Afterward, I read and studied Edmund Landau's *Calculus,* a text that in comparison was severe in its lack of compromise. Each page contained what to my eyes seemed a jungle of symbols and only a few words. The calculus, said the book, was as impenetrable as steel and hard as death. I was intrigued enough by the book to read a few remarks about Landau in a German encyclopedia of science. An imposing and successful mathematician and a man of great culture, Landau had been forced by the Nazis to flee Germany in the late 1930s. A picture taken in 1943 shows him lecturing in England, a lost look in his eyes. Yet it was Landau who had delivered the memorial address on the occasion of Dedekind's death in 1916.

1916? Somehow Dedekind *had* tottered into the twentieth century, one of those affecting figures who outlive their time, their place, and the circle of their friends. After all, Dedekind's intellectual roots lay in the eighteenth century. He had been a student of the great Karl Friedrich Gauss—his last. The terrifying old man had read and approved his Ph.D. thesis, and Gauss had been born in 1777! Every now and then I take down my copy of Landau's *Calculus* and let the heavy pages fall through my fingers. I seem to be listening for something.

A gabble of European ghosts.

appendix

A Cut
Corresponding to
No Rational Number

Here is the proof. Remember that there is no *rational* number r such that $r^2 = 2$. Now sever the rational numbers into two classes A and B, such that every number in A, when squared, is less than 2 and every number in B, when squared, greater. On the one side are the numbers in A, like firecrackers popping off when squared; no matter how they pop these numbers *never* pop past 2. On the other side are the numbers in B, like firecrackers popping off when squared; no matter how they pop these numbers *always* pop past 2.

Every number in A is less than any number in B, and so A and B determine a cut. The proof that this cut is produced by no rational number proceeds by driving an assumption into a contradiction. Suppose that there does exist a rational number r producing this cut. *Suppose it for the sake of argument.* By definition, r is larger than any *other* element in A. A number r' larger than r thus finds itself out in the badlands in B. When squared, r' *must,* by the definition of B, return a number greater than 2. However, no matter the candidate r, it is easy to produce a number r' greater yet whose square is *less* than 2.

How easy? *This* easy: Let $c = 2 - r^2$ and take r' to be $r + c/4$. Then r'^2 is simply $r + c/4$ times itself, that is, $r^2 + rc/2 + c^2/16$; but $r^2 + rc/2 + c^2/16$ is less than or at most equal to $r^2 + r^2c/2 + c^2/16$. After all, $r^2c/2$ is a bigger number than $rc/2$. But, look now, at this: $r^2 + r^2c/2 + c^2/16$ is just $2 - (7/16)c^2$, the identity emerging when the indicated operations in $r^2 + r^2c/2 + c^2/16$ are carried out. (How? By allowing 16 to be the greatest common denominator of a single fraction.) But $2 - (7/16)c^2$ is flatly less than 2 and r'^2 is no greater than $2 - (7/16)c^2$, hence *less* than 2 as well. The assumption admitted for the sake of argument, however, that r' is greater than r, implies that r'^2 is *greater* than 2. A contradiction has been reached. The party and the proof are over. It cannot be said that this is a very pretty argument, but it works.

chapter 9

Real
World
Rising

IN THE BEGINNING, THE NATURAL NUMBERS, 1, 2, 3, 4, ... THEN 0 AND the negative numbers. Next, the rational numbers, or fractions. And finally the irrational numbers. I have not said what the irrational numbers *are,* only that the real number system obeys Dedekind's axiom. Like members of a goofy lodge, the other numbers express their identities unselfconsciously, but the square root of 2? It has come into existence as the result of an assumption; it stands to the other numbers in a certain relationship; when multiplied by itself it yields the number 2. But after all is said and done the thing seems determined entirely by the relationships it entertains, rather like certain society figures who, in a phrase that is now well known for being well known, are well known for being well known.

A rational number or fraction, it is worthwhile to recall, enjoys a double identity, one that is on many occasions useful, as double identities often are. The number 1/2, for example, may be written in decimal notation as 0.5 and the number 15/28 as 0.53571428571428. Now the square root of 2 may *also* be written in decimal notation, for a start as 1.414. The notation serves to restore the irrational numbers to a certain community of numbers, for, in form, 1.414 and 0.53571428571428 appear to be objects of roughly the same kind. To the extent that decimal notation serves this psychological purpose, no harm is done. But the decimal expansion of a *rational* number—the numbers after the decimal point—is either finite, as in the case of 0.5, or doomed to repeat itself after a period, and so appears among the numbers as one of those tiresome ghosts returning every Halloween to the same fireplace, where they may be found rubbing their hands and looking mournful and making clanking sounds. In the decimal expansion of 15/28, the sequence 571428 occurs over and over again, clanking away.

The contrast to the irrational numbers is striking. The decimal expansion of an irrational number *never* repeats itself. Instead, the expansion trails off into the far future, each of its digits something of a surprise, the result a unique and infinitely long object with little by way of pattern or plan to ease the understanding. The square root of 2 is 1.414, and beyond that 1.4142, and beyond that 1.414212552 . . . ; from what has gone before, there is no telling what is to come. The digits expressing this number are unpredictable, random, unique, solitary, infinite, and unfathomable. They retain an element of unavoidable mystery. Like the human soul, an irrational number is only partly known, and however more is known of either there is always infinitely more to know.

Whatever the ultimate identity of the irrational numbers, what is known about them is of less importance than what is known of the great system in which they are embedded.

That system is *severable*. Dedekind's axiom is in force, flooding the numbers with light, flushing the irrationals from the shadows. Addition, subtraction, multiplication, and division, the immemorial operations of childhood, are entirely enabled; thus enabled, they allow the irrational numbers to function *as* numbers: $\sqrt{3} + \sqrt{12} = 3\sqrt{3}$, *be-*

cause the square root of 12 may be written as $\sqrt{4 \times 3}$ and then as $2\sqrt{3}$, making three of those square roots in all.

The system is *ordered*. Any number if it is not equal to 0 is either greater than 0 or less than 0. It is a system in which every number finds its place and there is a place for every number.

And the system is *complete*. There are no gaps to be filled. Any cut among the numbers falls like the stroke of an ax upon a single number. Positive numbers have roots within the system. The strange black nothingness that opened up among the rational numbers is gone. Incommensurable magnitudes are no longer incommensurable. The correspondence between the geometric line and the real numbers is perfect and unblemished.

At Last Another Country

Although each step is like the last, in the end the landscape changes. Descartes achieved a part of his vision of a number-infused, a number-*intoxicated,* world without being able in any way to offer a precise account of the real numbers; he was informed only by a prophetic sense that the details could be developed later. It is now later. The details are all of them available and need only be put in place. It is by means of a Cartesian coordinate system that the mathematician means to represent the real world: the real numbers now make their own circumspect but inescapable appearance in a coordinate system *via* an enrichment of the system's coordinate axes. The choice of unit and origin proceeds as before, but points on the line are mapped to real numbers and points in space to pairs of real numbers. These last steps having been taken, the construction of this system is complete and a representation of the world emerges: luminous, jeweled, black against a gray sky.

Yet there is in this structure still something sterile. A Cartesian coordinate system as it stands is a way in which to represent space in two dimensions, the x-axis stretching to the right and left, the y-axis reaching upward and downward, both x- and y-axes understood intuitively as representations of spatial distance or expanse. But the fantastic leading idea that animates the calculus has been to fashion from the real numbers and a sparse geometrical setting a representation of the *real* world in which things come into being, grow, and then decay. In the coordinate world, nothing lives and nothing breathes and there are surely no dancing monkeys in the heavy hanging trees. A Cartesian coordinate system represents space but *not* time, and without time there is no life.

Just before an utterly new landscape reveals itself in all its loveliness, there is often a moment of confusion, the eye requiring an inward adjustment before the scene comes into focus. It is by means of just such an adjustment in attitude that a Cartesian coordinate system acquires its living aspect. The geometric line is the ancient emblem of space, time, movement, and consciousness. The coordinate axes are themselves straight lines. The x-axis until now representing space, suppose instead that it represents time. Suppose instead that it represents time? Let it be so. The result is a Cartesian coordinate system in which time and space are represented and represented in one comprehensive structure and represented *simultaneously*.

No new steps have been taken, but things that are new may now be seen. Each point in *this* system is addressed by a coordinate indicating space and a coordinate indicating time. Where before, the system could only affirm that the legendary Leper's Depot was located at an address some distance along the x-axis and some distance along the y-axis, it says now that Leper's Depot is some distance along the y-axis *and* that *there* it is twelve o'clock. One spatial dimension has been lost, this to be made up later in advanced calculus, but a temporal dimension has been found. The difference between one point and another comes to signify a change in position *and* a change in time so that running through the temporal axis, the mathematician may see Leper's Depot in the flushed morning of its first success or in the enervating evening of its decline, great moths banging against the streetlamps and a cool wind blowing from the soundless prairie beyond the town. And in the end, what is life itself but a record of *where* something is and *when* it has been there?

The story is not complete: like so many real stories, it is *incompletable*. But as a real world arises from an act of the imagination, a last haunting bit of landscape loveliness comes into view. A Cartesian coordinate system embodies the assumption that in the real number system a *single* mathematical structure suffices to describe *both* time *and* space. The discovery that superficially different things or properties may be bound by a common description is often liberating, a truce in the great war between men and women being occasionally engendered by the abashed admission that they (or we) are after all human beings. Representation by means of the real numbers does not make time into space or space into time. The very idea is incoherent. But the representation suffices to show that contrary to every expectation time and space do have a common description and so find their place in a still more general scheme, one that reveals that on some very deep level the differences between time and space are of less importance than the fact that the real numbers represent them both.

The real world has now been interpreted within the real numbers, the first step in a mighty synthesis forged.

The Man Who Would Say No

Almost always, someone somewhere is forever saying no. Leopold Kronecker now enters into my story. Like Bishop Berkeley before him, Kronecker resides in the history of the calculus as one of the stern, incorruptible naysayers, a man unyielding in his intellectual demands and so something of a pest. Born in 1823 to a prosperous Prussian family, Kronecker was a German Jew and thus by virtue of a strange inversion of history's causal axis haunted by the future instead of the past. He was as a young man bright, capable, clever, and short. The center of his later mathematical life was in Berlin, that dark imperial city; and he figures forever as a discordant member of a quartet of powerful mathematicians—the others, Karl Weierstrass, Richard Dedekind, and Georg Cantor—who constructed the foundations of the calculus, and so completed the cathedral, in the second half of the nineteenth century. Cantor, Dedekind, and Weierstrass in some sense agreed on an *enlargement* of mathematics so that it included the irrational numbers. Not so Kronecker. Theirs were personalities of an instinctive intellectual generosity; his was a personality in which the margins of the permissible were retracted. This is not a moral distinction or even a rebuke. Mathematics no less than any other serious subject needs men who can say no.

Kronecker spent his twenties occupied by *geschäft* and was successful enough in business that he was able to devote the rest of his life to mathematics in comfort. I should like to pass out some snapshots. Here is Kronecker in early middle age, an elegant, dapper man, his monogrammed French cuffs turned out, instructing his English bootmaker to raise the heel of his shoe. And here is Kronecker bending his head low to confer with the headwaiter at a better Berlin restaurant, inquiring whether the white asparagus (*der Spargel*) has arrived yet. *Aber sicher, Herr Professor Doktor,* says the headwaiter. And here is one of Kronecker rising at a conference to attack Karl Weierstrass. You can see that Kronecker is pounding at the floor with his walking stick. Evidently the Chair has not yet recognized him. The rumpled and avuncular Weierstrass, a sheaf of papers clutched in his pudgy fingers—he has been giving a paper on his recent work on trigonometric series—is looking somewhat embarrassed. And here is one of Kronecker receiving the news that his academic enemy Georg Cantor, the creator of set theory, has retired to a mental institution. *Aber natürlich,* Kronecker is saying, *da gehört er hin.* That is where he belongs.

But the essential Kronecker resides entirely within an attitude, a tone of voice.

"The natural numbers," Kronecker is forever saying to the winds of history, "are the only numbers that assuredly exist. They are," he adds solemnly, "given to us by the Almighty. Everything else is the work of man."

And? the winds of history respond.

"There is no *and*. Nothing other than the natural numbers has a certain existence."

You've got to be kidding, Leo.

"No negative numbers."

Hey, that's cool, say the winds of history. *We never believed in them anyway.*

"No fractions."

Far out.

"No negative fractions."

You're on, say the winds of history, prepared as ever to favor self-confidence.

"And above all, no irrational numbers. The square root cannot be expressed as the ratio of two integers. That means it does not exist."

But, Leo, listen, Leo, how about them distances? You know, on the line and all?

"You cannot measure certain distances with a number. What of it?"

You can do that, Leo? I mean, I thought mathematics was like this it's got to be this way or else sort of subject.

"You were wrong."

Bracing, you say? And deliciously iconoclastic, speaking thus to the winds of history? Yes, of course. Like a cold clean shower. Deny the existence of the irrational numbers and there is no problem.

But no calculus either.

chapter 10

Forever Familiar, Forever Unknown

THE GREAT STAGE HAS BEEN SET. TIME AND SPACE THE MATHEMATICIAN has represented by means of a Cartesian coordinate system. It is yet a curiously *empty* stage. Like two lost and lonely railroad tracks, the co-ordinate axes stretch to the north and to the south and to the past and to the future. The Cartesian plane itself is suffused with a strange and somber silence. The air is absolutely still. Nothing happens.

But in *our* world, there is generation and corruption and dazzling phosphorescent decay; things have a boisterous beginning, they come to an unalterable end, and these simple human facts—the facts of *change*—mathematics is yet unable to reflect. As with so many other fundamental concepts, there is no saying what change *is,* the formula or form of words *change is* defined either with a knowing shrug or some verbal flourish patently the same as the concept under analysis. Change is *growth*. But growth is *transformation*. And transformations are *changes*. In talking about change, philosophers have made use of a vocabulary essentially no different from that engagingly presented by the ancient Greeks. There is the dusky river from which a dripping Heraclitus emerged, convinced improbably that he could never step into the same river twice. There are the paradoxes of Zeno, mad, bad, and dangerous to know. And there is not much else. But the analysis of change has been the mathematician's stock in trade at least since the seventeenth century. It is change that is the concern of the calculus and the inscription of change that brings a coordinate system to vibrant life; and if the mathematician cannot define change he *can* sort out its characteristic forms, the ways in which it appears in this, our crowded world.

We all of us live within hearing of the muted or monstrous sounds of a great clock, now ticking faster, now slower, but inevitably and inexorably *ticking,* and it is by reference to the clock that we measure the terrible and depressing changes in our own bodies, stomach expanding, skin sagging, arches falling, the story inconveniently reflected in the morning mirror, where a suspiciously familiar impostor apparently holds court. Such somber talk has at least the instructive effect of suggesting that change in something—change in *anything*—takes place against an assumed background in which time itself is changing, sagging skin sagging with respect to the time then and the time now, although how it is that time might change without some further standard of time to measure *that* is another mystery of the sort with which mathematics is strangely replete.

It is the physician who studies the corruption of the flesh, the changes in the body; the novelist, his spiritual brother, studies social life, corruption in the large. Their work proceeds by means of an accumulation of detail, the doctor and the novelist alike men with a professional interest in getting and hoarding facts.

The mathematician's interests lie elsewhere; his method is different.

A round red soccer ball is thrown into the air and after a while it comes back to earth. In answer to the question *what happened?* (or, perhaps, *what really happened?*), novelists know that the event and the way that it is described—its *story*—are inextricably linked, the event, or events, becoming what they are and gaining a sense of their identity from the stories told about them and the stories gaining their point and their purpose from the events that they describe.

Observing the ball going up and coming down, the novelist is irresistibly inclined to *amplify* the details. The ball going up, hanging there, hung, the sun sparkling in the sky, the air shimmering, the ball turning in all that glimmering gold, its white seam distinct against the red leather, the ball heavy now, rotating languidly in the calm clear air, falling faster downward, down, the ground and the grass, dew on the lawn, the ball bouncing as it hits, and then bouncing again, a final sodden thump, there, over there, a puppy on the lawn, the smell of lemon blossoms, a young girl in cut-offs, her red lips arched together in concentration.

Now, human beings are intuitive novelists, one reason that men who would never dream of attempting neurosurgery believe that they, too, have a story to tell and feel urged to impart it to strangers on social occasions. We are a hopelessly inquisitive species, and the idea that knowing something is a matter simply of knowing ever more about some things is a part of our immemorial inheritance. No doubt the urge to hoard facts is stapled to our genes. But observing the same ball going up and coming down, the mathematician is inclined to *minimize* the details, his intellectual movement retrograde to that of the novelist or physician. This turning away from various particulars requires great discipline. Revisiting the facts, the mathematician must *resist* the tug of those rich, very voluptuous descriptions of reality that the novelist or the physician might favor, dismissing them curtly in favor of two austere abstractions—*change in position* and *change in time*. Under the mathematician's hands, the world contracts, but it becomes more lucid.

In the context of the calculus, change in position is change that takes place along the spatial axis of a coordinate system, but the spatial axis itself may serve as a stand-in for *any* change that is made measurable by the real numbers, so that change in position functions as a large, a fabulously *general* concept, one standing in for change in *something*. It is the miracle of the calculus that change in something and change in time may be coordinated by means of the vastly greater abstraction of a *function*, purely an intellectual object, the key to the calculus, the key, in fact, to mathematics, and one of the great imponderables that like certain movie stars is forever familiar but forever unknown.

The Mathematicians' Lounge

A function indicates a relationship in prospect, and so belongs to a family of concepts. Relationship, as in *related;* relationship, as in *connected, corresponding to* or *caused by, united* or *bound together;* relationship, as in *linked* or *yoked, coupled* or *conjoined, associated* or *allied*. Relationship, as in *dependent,* indeed, relationship, as in *function of,* at which point the moving conceptual point may be seen revolving around the perimeter of a circle.

The textbooks offer an official version. A function, those thousand bright and brittle textbooks say, is a rule that assigns to each element in a set A a unique element in a set B. On the left are the elements in A, on the right, the elements in B. The function acts to *pick one* in A and *assign* it uniquely to one in B. This definition is current in the mathematicians' lounge, where the mathematicians gather after class, and where it is always four on a gray Friday afternoon, the rain just beginning to streak the sooty windows. The image of a function thus evoked suggests one of those ghastly preadolescent dances in which sullen boys are lined up along one side of the ineffaceably smelly gymnasium and preening girls along the other, an energetic social science teacher seizing one of the hideously embarrassed boys and, after dragging him by the lapels of his stiff sports jacket, depositing him in front of a pleased but stout and red-faced young girl: *Gregory, you dance with Jessica here*. The homely tableau succeeds in spite of itself. The sets A and B are represented by boys on the one hand, girls on the other, and the function itself by the Czar's dancing mistress, mysteriously transposed to suburban Teaneck, New Jersey, and acting energetically to *pick* a boy and *assign* him to a girl.

The functions of the calculus are mathematical functions, of course; they serve to bind real numbers to Significant Others that are real numbers in turn. The function f sends every real number to its square. In the calculus, the function's action is indicated by letting the function go to work on a number—2, for example. The notation $f(2)$, which is read aloud as f *of* 2, represents the result of *applying* the function *to* the number. Since f is in the squaring business, $f(2) = 4$; this last read f *of* 2 *equals* 4, or f *of* 2 *is* 4, or even *at* 2, f *is* 4. The function goes on to associate 3 with 9, the square root of 2 with 2, and −312 with 97,344 (this because −312 × −312 = 97,344). *Yielding, denoting, designating, indicating, giving, determining,* or *describing* is what the function *does* and what it *yields, denotes, designates, indicates, gives, determines,* or *describes* is the *result* of applying the function *to* a number. The number being acted upon is the function's *argument*; the result of all that action, its *value*.

Let f take care of squaring things. The function g *adds* the number 2 to any number, assigning 3 to 1, 5 to 3, 121 to 119, and 2 generally to x so that $g(x) = x + 2$—the entire number system, under the influence of this function, taking a hitching step forward by two places. Yet another function h squares a given number and then adds the number 2 to the result, merging the action of the functions that have just been defined. *This* function sends 4 to 18, 5 to 27, and x to $x^2 + 2$.

Functions, I have said, are rules or regulations, precepts or principles, and up till now they have remained in the verbal background, their influence revealed by their results. The rules may themselves be expressed in the very notation in which functions have from the first figured. The function f answers to a simple rule: *take a number and square it*. And whatever the limitation of ordinary language, *these* words get the job done. The rule goes over gracefully to symbols: f *of* x is x^2, with the *of* carrying enough intonational emphasis to suggest f actually applied to x and ferrying the results over to x^2. The result of applying f to a number x, the symbols say—whatever x is, whatever *it* happens to be—is that same number squared; in place of ordinary language, the rule is seen now through the dark glass of a mathematical variable. The same effect may be conveyed by the *equation* $y = x^2$. The equation is couched in terms of *two* carelessly intimate variables, x and y, the value of y *determined* by the value of x. Given that x is 4, y *must* be 16. For any value of the variable x—any number that ultimately stands in for x—the equation obligingly returns a new number y. This is just what a function does, so that y simply *is* the result of applying f to x.

Yo, f of x is x^2 and *Yo again*, $f(x) = x^2$ or *Yo, at x, f is* x^2 or even *Yo, I'm talking to you*, $y = x^2$, the symbols in the end all saying precisely the same thing.

The Iceman Cometh

Mathematics is one thing, but logic is another, a discipline of symbols set in snow. To the world at large, the logician, with his sea-green eyes and never-smiling mouth, appears as just another arid academic; but among the heartbruised mathematicians, he is the Iceman, always and forever. The air chills when he enters the mathematicians' lounge. His message is always the same, and it is always one of reproof.

A number is given and then squared. The number 2 *becomes* the number 4, the number 16 *becomes* the number 256, and I have stressed the *becoming* in order to include the functions in the category of things that stage-manage creation and so participate in the mystery of the *demiurge*. Skeptics and students alike may wonder what it all means, or whether it means anything at all, their sense of dissatisfaction only half answered by the familiar litany from the lounge. *A function is a rule which pairs elements in one set with elements in another set.* When the Iceman with a fierce snort of derision dismisses these declarations as "utterly senseless"—the very words of the logician Andrzej Mostowski—he has the crowd in the gallery on *his* side. Rules? Relationships? A rule *doing* the pairing? *Hey, gimme a break.* And perhaps the trouble is something so simple as this: we can see what a function *does* (assigns one number to another): we can see how it is *defined* (in symbols, by a form of words): we cannot see what a function *is*.

The request for palpability is an old one, especially in mathematics, and it expresses an old frustration. Numbers, lines, points, and even the lowly sets of the now-forgotten new mathematics, are objects of an inescapable *shiftiness*—the more scrutinized, the less seen. And so it would seem with functions—so it *does* seem. Still, a function is a creature with multiple identities, and however much it may keep its inner self concealed in the cloak of that business about rules, it yet has an existence in the world of action, where it busies itself giving directions to the numbers. And with what effect! Pair by pair, they troop into consciousness, those drilled and docile numbers, present themselves for inspection, and troop away, an army of black and doubled ants, an endless moving *list:* ..., $\{-2, 4\}$, $\{-1, 1\}$, $\{0, 0\}$, $\{2, 4\}$, $\{3, 9\}$, $\{4, 16\}$, ..., to keep to the case of a function that squares each and every number. Functions? Let their inner selves stay hidden shyly in the shadows. The concept of a list is out there in the sunshine and so serves as a point of palpability, a place where the functions acquire an identity in the real world.

Wait a minute. You're talking about a list?

I am talking for the Iceman and *he* is talking about a list. It is clear and cold, that list, a thing that can be seen and sensed and so a thing that occupies a portion of the blessed space reserved for the indubitable. The mathematician's functions are obscure, numbers are abstract, points occupy position without place, *who knows* the secrets that the sets conceal? But a list—that is something that is *there.*

Well, yes, what a function does is clearer than what it is.

Stripping an imaginary wad of fat from his sides with a harsh downward motion of his fingertips, the Iceman indicates a mental motion—a *flensing* of intellectual blubber, as if to purge the last of the mathematician's illusions. What a function does *is* what it is.

And in one of those queer inexplicable moments, the lounge lizards hear all at once the Iceman's voice blend with a dozen different voices from a dozen different disciplines, the *strange identities* similar in their simplicity: the mental *is* the physical, the personal *is* the political, the mind *is* the brain, the soul *is* the body, the possible *is* the actual, the function *is* the list.

After that, no one speaks. No one, in fact, breathes, least of all the mathematicians.

A Farewell to Strange Identities

The cultivation of a concept resembles the sculptor's efforts to see an essential shape hidden within a great looming block of stone. Step by step, the sculptor hacks and chips until the blistered block yields up its marmoreal nude or its poised and minatory archer. The Iceman's definition of a function as a set of ordered pairs—that list in *his* language—is the work of the twentieth century, and yet functions entered into mathematical thought three hundred years before. Leibnitz, who named everything in mathematics, named *them;* and they are, those tense and troubled functions, there in the *Principia*. The years were required to chip away at the association between functions and rules, between functions and the various verbal flourishes or formulas by which they were defined. The Iceman's function is *not* the rule *nor* the formula, *not* the process *nor* the procedure. Emerging finally from the stone in which it had been entombed, the function appears naked as a pure form of association, an undertaking between numbers and between numbers alone, with no intervening form of words that define or defile them. *The essence of a relationship is the things that are related.*

This last is a strange and a powerful and even a haunting assertion. I have heard it from the Iceman's lips in a hundred different lectures and I have repeated it myself in a hundred lectures more. But now a secret must be imparted. In the mathematicians' lounge, the mathematicians hear the Iceman out, and in their lectures they even go so far as to quote what he has to say; but when he has gone, leaving behind only a hyperborean chill, they repair to the old, comfortable incantations: a function is a rule that assigns to each element in a set *A* a unique element in a set *B*.

They repair to the old familiar formulas. And so do *I*.

I suppose that this defiant disclosure calls for an explanation, but I have nothing better to offer than the observation that the world deals harshly with those who trifle with memory and desire: in the end, the materialist is rebuked by the eyes of a cat; and if this is not the sort of thing calculated to move the Iceman it is calculated to move *me*. I prefer to see the functions remain charged with their mystery. There is to the very idea of a function some intrinsic residue of human energy or agency without which the concept is incomplete and insufficient. A list or a set of ordered pairs may express or represent the action of a function; but as the soul is *more* than the human face or figure in which it lies hidden, so a mathematical function is more than the list through which it is revealed. The list belongs to the palpable universe of things; the function, to the world of energy and action and intention, the functions extending themselves over the numbers in response to electric affinities that we, the human actors in the drama, can barely begin to see or even sense.

So much for purple prose, the stuff beginning to wear, even on me; but if the mathematicians in the lounge are embarrassed at my defense of their definition, they do not let on, one of them murmuring *attaboy, Dave,* just after I have delivered myself of the last purple patch.

Faces of the Functions

The characteristic objects of the calculus are not numbers, neither are they places, points, times, or symbols; neither are they coordinates or coordinate systems. At the damp, dark, and bat-filled place whence functions arise, there the calculus finds its quintessential tools, its indispensable objects.

It is worthwhile to recall that two real number lines are entombed within the crossed and outstretched arms of a Cartesian coordinate system. Left to their own devices, they do nothing beyond what number lines generally do, which is generally to exhibit the real numbers. A Cartesian coordinate system understood simply as a geometrical object—an ornate mathematical map—is curiously inert; it represents time and it represents space, but it represents both time and space without bringing about a coordination between time and space and so functions as an arena rather like a Beckett play in which something is always about to happen.

A function taking real numbers to real numbers, on the other hand—a *real-valued function,* as it is known among pros—is all nervous energy: give the thing a real number and *whump* back comes a real number. Up to now, functions have made an abstract appearance, rather like pulses of dry lightning flickering in the humid night sky. That abundant energy might as well be transferred directly to a Cartesian coordinate system. Functions that take real numbers to real numbers may also serve to map one axis of a Cartesian coordinate system onto the other and so bring about a coordination between time and space. They now are written as $f(t)$ rather than $f(x)$, a change in notation signifying a connection achieved between time, represented by t, and space, represented now by $f(t)$.

The application of a function to a Cartesian coordinate system may be seen as well as specified in words, the effect entirely enchanting as whenever a series of directions—*follow the river and then you take a left at the Old Mill Road*—turns out to reveal a vivid and startling landscape, the featureless words leading to a feature-filled world, a blazing panorama, the high mountains in spring. Let me return to an old standby, the function that takes any real number and carelessly squares it—$f(t) = t^2$.

Where is it coming from? The real numbers.

Where is it going to? The positive real numbers.

Representative arguments and values may be arrayed in an obvious way as a table whose entries are pairs of numbers:

t	$f(t)$
-3	9
-2	4
-1	1
0	0
1	1
2	4
3	9
4	16

$$f(t) = t^2$$

But pairs of numbers also mark *points* in a Cartesian coordinate system, with the pair <2, 4> in the table corresponding to the point <2, 4> on the plane. Let those points be marked, at least for corresponding pairs in the table; they serve as isolated starbursts, explosive puffs, tantalizing spectral dots.

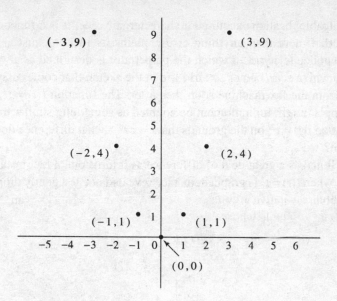

Points corresponding to $f(t) = t^2$

Connected in the obvious way by a curved line the dots organize themselves to reveal

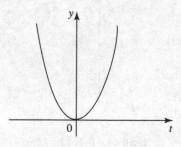

Face of $f(t) = t^2$

a *curve,* a connected, sensuous shape, a coherent object, something on which the senses can gratefully fasten, in this case, a gracefully dipping parabola.

A function, spun out of words and symbols and so something

impalpable, has now acquired a characteristic face. It is a function's face that more than anything else explains its identity, just as the terse police teletype, in which the perpetrator is described as a white male with a scar, comes to lurid life in the sketch that comes clattering from the fax machine soon thereafter. The function $f(t) = t^3$, for example, might for a moment be counted as something similar to the function $f(t) = t^2$, on the grounds that t^2 or t^3—what difference does it make?

It makes a great deal of difference as it turns out, a fact instantly seen when $f(t) = t^3$ is graphed, its face revealed not as a gently dipping parabola

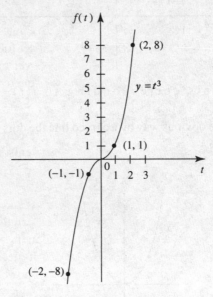

Face of $f(t)=t^3$

but as something entirely different, rather a Chinesed curve, inscrutably passing through the origin on its way south from points unknown in the north.

And there is more. With that face in focus, a new object arises in answer to the question—*a function?* What *is* it? Mathematicians may well insist on the important distinction between a function and its face; let them, by all means, insist. The curve may not *be* a function, which is—the mathematicians are right in this—a correspondence between numbers brought about by a rule or regularity; but neither are the eyes the same thing as the soul. They are the soul's window and serve to let light into darkness.

So it is with curves and functions.

Only What? Only Connect

It is now a part of human consciousness, the concept of a function, and thus an idea that divides things into before and after.

Before, the world is a world of things and individuals, a mean wind blowing across the mutinous sea, the dining hall set for dinner, a few musicians playing a mournful tune, over and over again, a dozen heavy-set men with beetle-browed eyes entering the room and taking their accustomed places, the stout archduke, walking with a limp owing to a savage attack of gout, sighing heavily as he takes his place at the head of the table. Grace said in horribly guttural German. Huge platters covered with slices of pork, bloody slabs of venison, a haunch of beef, the men eat furiously, gaffling and snorting.

No one says a word.

But *after,* the world is a world of associations and affinities, relatives and relationships, ten thousand flickering candles reflected in the polished parquetry, the warm light catching on the silver sconces, the Aubusson tapestries glowing. It glitters and it gleams, that light, aureate and ochroid, red and rust, courtiers with powdered wigs entering the great palace hall and drifting delicately through the crowd, bowing low to talk and listen and then talk again. The red-robed cardinal walking in stately and dignified grace, holding a perfumed handkerchief between his thumb and forefinger, surveying the scene with hooded eyes, nodding to members of the court and to the musicians and to his green-eyed mistress by the stairs. Music in the air, warm currents carrying the soft sounds. A woman with dark lustrous hair lifting a fan to hide her face, lowering it to reveal her eyes and the convex top of her rouged cheeks, the gesture multiplied in the polished mirrors, a marquis inclining himself slightly from the waist, turning his feet outward, lifting the delicately bunched tips of his fingers gently to his lips.

Only what? Only connect.

chapter 11

Some
Famous
Functions

THE ARABIC HISTORIAN AL-ISHAQUI TELLS THIS STORY: WISHING TO
reward a faithful retainer for his service, the Caliph al-Mamun invited
the old man to make a wish. Time collected itself into a shimmer and
then for a moment stopped as the retainer contemplated his desires.
Gold and silver he dismissed: mere things he held of no account. With
an enigmatic smile, he asked finally that a single grain of wheat be
placed on the first square of a chessboard and *doubled* every day thereafter on the successive squares until the last square of the board had
been filled. Sitting on his gold and purple silk cushions, the Caliph
closed his heavy-lidded eyes; he imagined the chessboard, whose
sixty-four squares divided his consciousness into contrasting regions
of darkness and light, and with a careless mental motion he filled

the board with harvest wheat. *Make it so,* he is reported rashly to have said.

Rashly? If a single word could be made to reveal its secrets, the whole of the calculus could be seen in the manifold of its meaning. The strange little story of the Caliph appears in puzzle books from time to time, and in the back pages of *Parade* magazine; and just recently I heard the story recounted yet again in the health club where I repair each afternoon, a dark-haired dark-eyed woman calling the Caliph into existence in order to explain the mysteries of compound interest to a rapt young man.

Like the Sorcerer's Apprentice, the Caliph calls into being forces that he does not understand and that he cannot control; but what lends to *his* story its particular power is its evocation of a consciousness that is not only confused but intellectually clouded as well.

Looking at a sunset along with his master, a dog sees what his master sees, at least to the extent that dog and master are seeing the same thing, the both of them observing something colored in the western sky. It is the ability to *name* things that confers the power finely to divide experience. The color of the sky at just the moment the sun disappears behind the far hills?—not red, not yellow surely, not rufous or rust, not rustlike, peach or peachy perhaps, the color of a peach that has been peeled, the bruised fruit like the bruised sky, the two of them in minutes losing their color along with their life. All this the dog cannot say and so cannot see. And yet as the Caliph al-Mamun discovered to his dismay, the rich and lustrous vocabulary by which we specify the sparkle in the evening sky—*that* fails us when we contemplate growth or decay or time's slow surge or swift-moving shimmer.

Mathematics is conceived in the fires of the real world, and the functions that bring twitching life to the calculus represent processes beyond the closed coffin of a coordinate system—things that take place in time and that take place in space. The maturation of an egg and the education of an undergraduate both represent processes; so does the circumnavigation of the moon, the beating of the human heart, and the dense shifting of the seas under the influence of the lunar tides. But not every functional relationship, it is important to recall, is a relationship between numbers. There is a world of relationships *beyond* the world of mathematics in which men and women pair off, children are raised, scores are settled, and fathers age and then die, and none of this is correlated with the real numbers. And not every relationship between numbers expresses an interesting mathematical function. Surrounded somewhere in Las Vegas by blue-haired women and hard-eyed men, a roulette wheel spins in the soulless night. It generates a string of random numbers. The *association* between those numbers and the time at which they are generated describes a matching of numbers to numbers and so a function. It is an association without structure and so without interest.

The functions of the calculus represent, they *embody,* the mathematician's canonical instruments for the depiction of change. They have up to now appeared as a collection. Mathematics is, in part, a rhapsodic subject, full of dark mystery; but it is also a discipline like comparative anatomy in which things are put in their mathematical places and places found for mathematical things.

The Polynomial Archipelago

The simplest functions are constant—These are functions of the form $f(x) = C$, for all x, meaning that whatever the number x, the function dutifully returns one and the same value: $f(x) = C$ as in $f(x) = 4$, or $f(x) = 32$, or $f(x) = \sqrt{2}$, or $f(x) = \pi$, or $f(x) =$ pretty much *whatever* so long as *whatever* is a real number and the *same* real number for every choice of x. The constant functions depict straight lines in space: those straight lines must be parallel to the x-axis by virtue of the fact that $f(x) = C$ is neither increasing nor decreasing. The face of these functions is regular and like that of an actor in his twenties devoid of character. Just look at the constant function $f(x) = C$, where C happens to be 2.

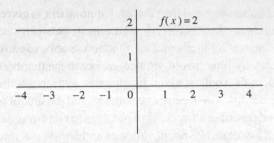

Constant function $f(x) = 2$

Traveling across the Cartesian coordinate system, the graph of this function is now and forever the same distance from the x-axis, always 2.

Power functions next—These are functions of the form $f(x) = x^a$, where a is an arbitrary but fixed and thus specific positive integer. The function $f(x) = x^2$ is a particular example, one sending 2 to 4, 3 to 9, 4 to 16, and so on up or down the long chain of numbers; but $f(x) = x^3$ is *also* a power function, one that maps 2 to 8 and 3 to 27; and so, for that matter, is $f(x) = x^{666}$, a function that demands that a number—*any* number—be multiplied by itself six hundred and sixty-six times; it is this function, no doubt, that the Devil will endeavor to invoke on Judgment Day, the image of the Evil One struggling with elementary mathematics irresistible somehow.

The root functions represent the power functions in reverse—These are functions of the form $f(x) = x^{1/n}$, denoting the nth root of x, a somewhat mystifying concept, those n roots apparently loitering about like gerbils; but if $n = 2$, the nth root of x is simply its old-fashioned square root, and if $n = 3$, its nth root is the number that when multiplied by itself three times is x itself. Let x be 36. The function $f(x) = x^{1/2}$ sends 36 to the square root of 36. One number—36—comes slithering up to the surface, and far below another number, its square root—6, as it happens—comes slithering up behind it. Third roots, fourth roots, and all roots expressed by positive integers behave in just the same way and express just the same relationship.

The constant, power, and root functions are included in the vast archipelago of polynomial functions, the term *polynomial* unaccountably suggesting Polynesia and so drawing a lovely but lunatic verbal association between the nature of these functions and their name.

Here is how they work. A stray set of real numbers is given—say, 5, 9, 32, and 6. And a single variable x, a solitary stranger. A polynomial function in the numbers 5, 9, 32, 6, *and* the variable x is any function that can be created by adding or multiplying these numbers and that variable: $f(x) = 5x^3 + 9x^2 - 32x + 6$ is an example; $f(x) = 32x^3 - 5x^2 + 6x + 9$, another. But $f(x) = 5x^3$ is *also* a polynomial function, the entirely prosaic result of multiplying x, whatever it happens to be, by itself three times and then multiplying *that* by 5. And so is the lowly linear function $f(x) = 16x + 10$.

Within the ordinary world of ordinary arithmetic, rational numbers are represented as the quotient of two integers. The operation of division enlarges the margins of the mathematically possible. Something is *done* to the numbers that exist; the doing over and the hocus-pocus said, *new* numbers have been made to appear. What holds for numbers holds again for functions, the ratio $f(x)/g(x)$ of two polynomial functions $f(x)$ and $g(x)$ coming into existence as a *rational function*—$h(x)$, to give it a name and fix it as a character. Let $f(x)$ be x^2 and $g(x)$ be $5x + 3$. The *rational* function $h(x)$ is the function $x^2/(5x + 3)$, $f(x)$ on the top, $g(x)$ on the bottom, just as the definition demands.[1]

If the polynomial functions constitute an archipelago, the algebraic functions constitute a greater, grander archipelago, one which comprises *any* function constructed from the polynomial functions by means of ordinary algebraic operations. In forming algebraic functions, polynomial functions may be added to one another, they may be subtracted from one another, their roots may be extracted, and any given polynomial function may be divided or multiplied by any other polynomial function. The rational functions are included in the class of algebraic functions. But so is the function $g(x) = (x^2 + 9)^{1/2}$, a rugged newcomer constructed from the polynomial function $f(x) = x^2 + 9$. The construction proceeds for each number x by first fixing the value of $f(x)$ and then extracting from that number its square root. At $x = 4$, $f(x)$ is 25 and $g(x)$ 5.

---

[1] A surprise quiz now follows. At $x = 3$, $h(x)$ is—what? Don't give up. Here is how it is done. The function $h(x)$ is $x^2/(5x + 3)$, right? Start then with top, which is x^2. If x is 3, x^2 must be 9. Go then to the bottom, which is $5x + 3$. If x is 3, $5x + 3$ must be 18. And $x^2/(5x + 3)$ must be 9/18, or 1/2.

In the destiny of these particular functions, the ancient operations of arithmetic may be seen operating, like familiar temple spirits at work in a new setting. Mathematics begins with the simple inscrutable natural numbers. At once playthings and atoms of the mental life, these numbers may be altered and combined by means of four fundamental operations: addition, multiplication, subtraction, and division. In the polynomial archipelago, functions have come to the surface. But the simple and elementary operations remain, the polynomial functions retaining in their very nature a lingering memory of the most basic of mathematical objects and the most basic of mathematical operations.

The Transcendental Atolls

The polynomial functions constitute an archipelago, and like the islands in an archipelago they are close to one another, the pale blue ocean waters bright, but not deep. The *transcendental functions,* by way of contrast, are *not* algebraic; they go beyond—they *transcend*—the algebraic operations responsible for the construction of the polynomial functions. They rise as isolated and volcanic atolls out in the open ocean where the waters are dark and where an object that is dropped would travel for years before hitting the ocean's sandy bottom.

First, there are functions that are exponential—$f(x) = a^x$. The power functions of old raise a variable to a specific power, as in x^2 or x^{34}. With the exponential functions, it is the other way around. The specific number sits at the bottom; the *variable* is up there in the cockpit, as in 2^x. This is a big, big difference. Power functions are polynomial. Exponential functions are not. The exponential functions are in the business of big-time growth. Say that $a = 2$. 2^5 is 32, 2^{10} is 1024, and 2^{20} is 1,048,576; at 2^{40} the result is too large to be conveniently calculated. Yet the exponents commanding the cockpit are doing nothing more than doubling in size.

Exponentials constitute a family of functions, a particular member of that family arising for each choice of a, the *parameter*. (Use of the phrase *one-parameter family of functions* is thought to be a mark of mathematical sophistication.) The function 2^x is exponential, as is

3^x. The various exponential functions are similar in shape and similar in spirit, *whatever* the number a.

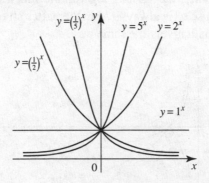

Family of exponential functions

They congregate at 1 when $x = 0$, because every number raised to 0 is 1. They all start off slowly. They all get big in a big hurry. And yet like members of a family they are all different. This suggests that there is in the concept of a function an unsuspected flexibility, a finely controlled capacity to vary by degrees.

Like members of a family, the exponential functions are *out there,* various as the numbers are various, and unified apparently by the circumstance that they all move things right along. My breezy introduction of the exponential functions serves only to palpate them on their padded shoulders; this must not be misconstrued as a definition. It makes sense to speak of 2^x, where x is a positive integer. Thus one has $2^1, 2^2, 2^3, 2^4$, and so on up the chain of command; but what of 2^π? The number 2 multiplied by itself π times? Whatever mathematicians might say, in this way lies madness. The calculus *does* provide a lucid definition of the exponential function; it reveals *how* the real numbers may be raised to the real numbers, but the definition comes later in the day and the exponential functions have made an appearance early on. In this regard, the calculus carries with it an unavoidable suggestion of logical lumpiness.

Logarithmic functions $f(x) = \log_a x$ next. The parameter a is again a fixed number, the *base* of the logarithm, the function serving to denote the number required to raise the function's base to its argument. For example, $\log_{10} (100)$ is 2—this because $10^2 = 100$. Logarithmic functions grow slowly and like the exponential functions resemble one another, the various logarithmic functions crossing the x-axis at 1 because a^0 is always 1, and thereafter all of them growing dolorously in qualitatively the same way.

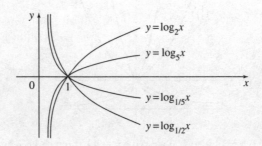

Family of logarithmic functions

The relationship between the exponential and logarithmic functions is one of shimmering *inversion,* a common pattern in mathematics, as when subtraction undoes what addition accomplishes, and a pattern that like a chronicle foretells the fundamental theorem of the calculus. The argument of the exponential function is the value of the logarithmic function. Say thus that $y = a^x$. This is an exponential function. Some number a is being raised to a real number x.

Version.

But wait: $x = \log_a y$. This is a logarithmic function. That real number x is the value of the logarithmic function at y.

Inversion.

At $a = 2$ and $x = 5$, $y = a^x = 32$. And $5 = \log_2 32$.

Version and *inversion.*

Logarithms taught in high school—the *Briggs logarithms*—take matters to the base 10. In computer science, where binary bits are basic, the requisite base is 2; but in mathematics the base of all bases is the transcendental number $e = 2.71828 \ldots$. It is a number of strange Sicilian dignity. The most natural definition of the logarithmic function, the definition that is the most *mathematical,* leads to the most natural definition of the exponential function, which in turn leads to the number e. The intellectual movement is one of a soapy wave washing forward to define the logarithmic function, washing backward to reveal the exponential function, and exposing in the back and forth motion the number e, the black jewel of the calculus. Once uncovered, e serves to unify, to *amalgamate,* all other exponential functions. The function $f(x) = e^x$ is itself an exponential function, one taking the number e to the power x. But the laws of logarithms permit *any* exponential function a^x to be rewritten in terms of the natural exponential function. Rewritten how? In the simplest possible way. A number serving as a logarithmic base carries the power to cloud men's minds: it can impersonate any other positive number a. The impersonation proceeds when the base is raised to the number's logarithm: $a = e^{\log_e a}$, to put the matter into the mathematician's suave symbols. But then, a^x is $(e^{\log_e a})^x$, or, what comes to the same thing, $e^{x \log_e a}$; and the extraordinary thing is that a^x has disappeared, its exponential voice entirely recovered by $e^{x \log_e a}$.

Does this act of notational unification transcend itself to affect an attitude and make a statement? It seems to suggest that beneath the various exponential processes there lies lurking some primordial exponential process. I do not know whether this is so. It is worth observing, however, that it is mathematics that makes the question possible; like those commanding exponents raising numbers to stratospheric heights, a small investment in symbolism returns large intellectual dividends so that in the end the study of the exponential function is itself a form of exponentiation.

Finally, there are the trigonometric functions—**sine**(x) and **cosine**(x), the hapless tangent brusquely defined as the ratio of the **sine** and **cosine** functions. In the trigonometry of blessed memory, the trigonometric functions are defined in terms of the sides and angles of a right triangle, opposite and adjacent sides of that triangle morosely mounting one another or a vagrant hypotenuse. The definitions come swimming up from memory, barely alive and gasping for air. Given an angle θ, Greek unaccountably the language of angles, the **sine** of θ is defined as the ratio of the side opposite to θ to the triangle's hypotenuse; its **cosine**, as the ratio of the side adjacent to θ to its hypotenuse.

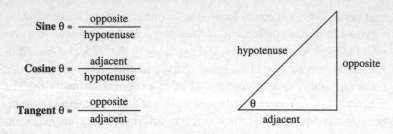

Sine θ = opposite / hypotenuse

Cosine θ = adjacent / hypotenuse

Tangent θ = opposite / adjacent

Trigonometry of blessed memory

In the calculus, degrees disappear; the concept of an angle is subordinated to the concept of a real number. Subordinated how? Subordinated thus: The circumference of any circle is given by the formula $C = 2\pi r$, where r is the circle's radius, a spoke going from the center of the circle to its perimeter, and π is a real number, a mysterious and ubiquitous constant. Let $r = 1$ and C returns as 2π. And 2π is a real number, like e among the transcendentals, but a number still and amenable to capture by a function.

A right triangle is now embedded within the unit circle, its legs extended so that they touch the circle's perimeter.

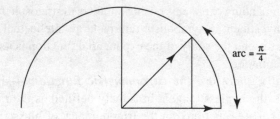

Unit circle with inscribed triangle

An angle is *defined* as the length of the arc that those kicking legs connect. The arcs represent real numbers—fractions of 2π, in fact. So, too, those angles, arcs serving to *represent* angles and vice versa. The old lost language of degrees is easily translated into the new language of the real numbers. An angle of 45 degrees subtends an arc covering one-eighth of the circle's circumference. One-eighth of 2π is $2\pi/8$ or $\pi/4$. Forty-five degrees corresponds, then, to the number $\pi/4$. An angle of

180 degrees corresponds to π itself, and an angle of 360 degrees, a leisurely full sweep around the unit circle, to 2π.

The **sine** and **cosine** functions may now be observed bursting buoyantly from their chrysalis to take up a new identity as circular trigonometric functions. The angles of old are represented by arcs or arc lengths. Those right triangles in which the trigonometric functions were imprisoned may like dry husks be allowed to disappear along with their degrees, leaving behind only their extended legs. With one leg fixed, the other is allowed to revolve around the unit circle, tracing a moving point whose coordinates are $<x, y>$. The **sine** of a given angle is defined as the ratio of y to r, its **cosine** as the ratio of x to r.

And the remarkable thing is that when that barely breathing right

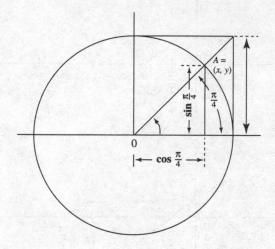

Circular trigonometric functions

triangle is allowed to reacquire, if only for a moment, its living shape, *this* definition coincides completely with the definition given long ago. Did the trigonometry of blessed memory define the **sine** of an angle as the ratio of the triangle's opposite side to its hypotenuse? It did. But then so does the definition just given, for y is the height of the triangle's opposite side and r the length of its hypotenuse. In mathematics, unlike life, nothing is ever lost forever or given up for good.

The trigonometric functions map real numbers to real numbers and belong thus to the large and noble family of real-valued functions. They come onto the scene by means of a cumbersome definition, one mixing memory (those right triangles) and desire (the need to incorporate all functions within the class of functions taking real numbers to real numbers); but their importance lies less in how they are defined than in what they are. After a few hitches or twitches, the polynomial functions settle down to behavior in which things simply get bigger or smaller. An example is $f(x) = x^2$. Just once, as the values of x approach 0, the graph of this function dips downward to alter its shape; thereafter it ascends solemnly like a helium-filled balloon.

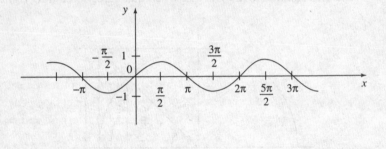

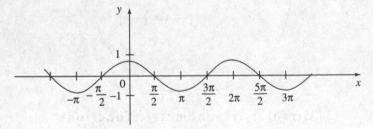

Faces of **sine**, **cosine** functions

The exponential functions embody all the manic energies of growth; but the trigonometric functions, *they* exhibit an utterly characteristic *periodic* face, repeating their behavior again and again and again. Periodic phenomena in nature range from menstrual cycles to the cycles of the moon, and the trigonometric functions embody in their regularity some essential rhythm of the universe. An invisible inner connection exists among the trigonometric functions, one revealed in various trigonometric identities, strange places where the trigonometric functions appear fluidly to exchange identities or to resolve themselves by means of simple arithmetic operations into unlikely numbers. At any real number x, for example, the square of its **sine** and the square of its **cosine** is always and inevitably 1: $\sin^2 x + \cos^2 x = 1$. Like the exponential functions, all of them subordinate in the end to one function, the trigonometric functions give the appearance of clutter and confusion masking some form of unity.

The constant functions, the great archipelago of polynomial functions, and beyond them the greater archipelago of algebraic functions, and far beyond *that,* separated from the polynomial functions by the mid-Atlantic trench, the transcendental functions just described—these are the *elementary* functions of the calculus.

A Cosmic Catalog

And this is how the elementary functions should be kept resident in memory, as a mathematical mandala, a collection of distinct faces, each function carving out from empty space a certain characteristic and descriptive shape, a way of describing and so a way of being.

The mandala is what may be seen; the functions by which it is generated represent the commanding analytical voices compelling the mandala into being. Like everything that may be seen and so remembered, the faces on the mandala are evanescent. It is the functions that retain the secrets of their identity. However many times the mandala is wiped clear, the sand of its creation brushed into oblivion, the functions themselves remain, and so the mathematician, like the monk, retains the ability to create the mandala anew.

The elementary functions represent an enlargement of the human capacity to notice the nuance in things. They might have offered to the Caliph a system of classification fine enough to accommodate the data of his experience. His consciousness clouded because lacking in powers of discrimination, he might have profited from the observation that doubling processes in nature may often be described by a simple *exponential* function—$f(t) = 2^t$. [1] It is *this* function that describes what

---

1. Time is now of the essence: x has become t.

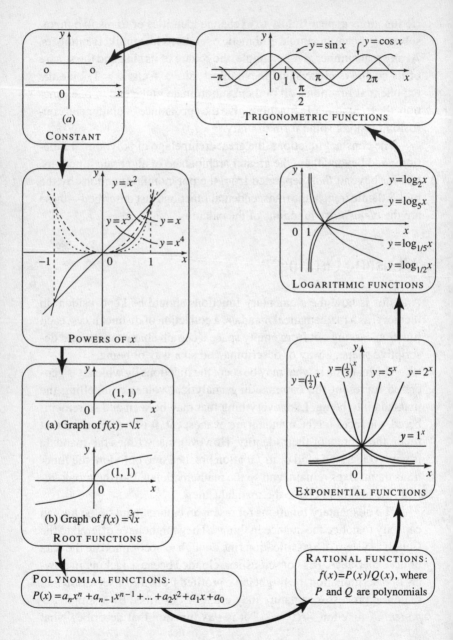

Mandala of elementary functions

takes place on the groaning chessboard. First, 2^0 is 1. There is on the first day one grain of wheat on the board. Next, 2^1 is 2. Fructifying in accordance with the Caliph's wish, the grains have doubled. Then, 2^2 is 4, the grains have again doubled, and so on to the point that the board breaks, the game ends, and a vast wind of desolation spreads over the kingdom. Left to his own devices, the Caliph is able only to guess incorrectly at the nature of growth. *The function $f(t) = 2^t$ precisely describes the process*. Left again to his own devices, the Caliph is able only to estimate incorrectly the number of grains required on the sixty-fourth day. *The function precisely describes the number*. It is 2^{64}.

Seen in one way, as, no doubt, they are most often seen, the elementary functions comprise a discouraging collection of odd formulas and wretched rules. But seen another way, as functions from time to space, they constitute a catalog of actual and possible processes, a small book in which the ways that things might happen are meticulously described. This is a remarkable idea—a book of processes, a kind of *cosmic catalog*. The mathematical mandala represents the way the functions are written on the wind as graphs or pictures; the catalog represents their inner nature, their analytical identity. The elementary functions embody the first great mathematical classification of the natural world to go beyond counting or Euclidean geometry and so constitute an audacious expansion in our own human power finely to divide the manifold of experience.

Do they *suffice*, the elementary functions? Is everything that can be described describable in elementary terms? Or are there processes that cannot be described by the elementary functions, as there are colors that cannot be described by the twenty or so color words in the English language? *Ahh*. Let me break in with a sigh. They do not suffice. The world contains more than elementary processes.

But this, too, is a lesson taught by the calculus.

Loitering at Infinity

The fire of mathematical creativity—who knows why it burns or where? Fifty years after Newton and Leibnitz saw their way through to the heart of the calculus, the center of mathematical thought passed to French-speaking mathematicians, and it is in the far-flung academy of their genius that the calculus became not only the principal instrument in the development of mathematical physics but the pathway to *analysis* itself, the subject in which the mathematician attends to the infinite. Leonhard Euler's great text *Introductio in analysin infinitorum* represents one of the post houses of civilization, places where the great tired horses bringing the news from far away are rested and exchanged for horses that are fresh.

Euler was born in Basel in 1707 and exhibited throughout his life a sense of the intellectual sublime somewhat at odds with his stolid Swiss citizenship. He was a man of the eighteenth century, the largest mathematical figure of his time. Accounts of his younger years—he received his degree from the University of Basel at the age of fifteen—suggest an intelligence that like the lilies of the field achieved its effects without effort and so inspired lesser men to the charge of superficiality.

The son of a minister and thus heir to the dour Swiss Calvinist tradition, Euler came early in life under the spell of the Bernoulli family, an extended clan (father, various brothers) marked by the shyly recessive strain of mathematical genius, and members of the family came under his spell in turn, recognizing his lucent intelligence and urging Euler's father to rescind his ill-considered demand that Euler devote himself to theology.

In 1727, Euler accepted a position at the St. Petersburg Academy; he was, no doubt, attracted by the warm liberal glow retroactively thrown as far afield as Switzerland by the court of Catherine the First. It is quite irresistible to imagine Euler thus, living in snowy St. Petersburg amidst violent and profane Russians, his attendance at court a matter always of moments of sheer terror, the primitives in charge of things *after* Catherine's death much given to whimsical executions and appalling cruelties, and through it all, stamping his feet in the fresh city snow, or attending to his enormous family—he was father to thirteen children—the papers, monographs, and books accumulating in piles, an idea seizing him in the afternoon like the tickle before the sneeze, fully fledged by dinner, the paper itself (and thus the sneeze) dealt with and done soon after.

It was Euler's splendid mathematical *profligacy* that marks him most as an eighteenth-century figure; he was cheerfully willing to direct his intelligence toward anything that interested him: puzzles, small problems, computational techniques, odd little theorems, major problems, research of every kind; he had a fantastic gift for formulas, seeing a secret world in symbols; he carried on an extensive correspondence; he amused himself with mathematical oddities, his intelligence functioning in that strange frictionless world which in the history of mathematics is inhabited only by Euler himself and in the history of music only by Mozart. It was Euler who saw and understood the fact that exponential and trigonometric functions are related, each definable by an appeal to the other. And it was Euler who discovered the most beautiful formula in all of mathematics: $e^{\pi i} + 1 = 0$, a mysterious and ineffable expression connecting the five most important numbers in the universe.

In 1740, Euler accepted an invitation from Frederick the Great to join the Berlin Academy. At court, he was unaccountably abused by Voltaire. He was unable to cut a clever figure. He seemed to have acquired a stammer. He lost the indulgence of his patron, an episode that indicates that even men with large-hearted sympathies may on occasion behave as blockheads. He returned to Russia in 1766, Catherine the Great having ascended the throne. He was received with kindness.

Euler lost his sight in old age and passed the last seventeen years of his life in darkness. His mathematical research did not slacken; the computational activities for which he was famous he now conducted mentally.

He sits there by the fire somewhere in St. Petersburg, an old man with vein-ruined hands, a shawl over his shoulders, dreaming in the darkness as all around him a Russian winter gathers, the darkness behind the darkness approaching.

chapter 12

Speed
of Sorts

SPEED IS ONE OF THE TWO CHIEF CONCEPTS OF THE CALCULUS. THE OTHER is area. They appear strangely unrelated, the mathematician's insistence that they be placed in proximity at first suggesting one of those by now clichéd paintings in which an attenuated guitar is improbably juxtaposed against an old-fashioned pair of reading spectacles. Appearances are misleading. Speed and area resemble two tightly twined concentric bands of white and yellow gold. This conceptual connection the fundamental theorem of the calculus reveals; but the revelation depends on—it *requires*—a reorganization of familiar ideas. The reorganization complete, speed and area acquire new identities or aspects, speed as the *derivative* of a function and area as its *integral*. These are technical terms, their definition for the moment deferred.

And yet syntax suggests, as syntax often does, something striking in the shape of things to come, the phrase *the derivative of a function* implying, whatever it might mean, a subordination of speed to the idea of a function. The system of concepts already in place is unrelenting in its demands.

Moving Along

Whatever else it may be, speed is something felt directly by a living human body, as when straddling that thundering Harley-Davidson, the one my family assures me in real life that I will purchase over their dead bodies, I roar down a desert highway, tires humming on the road, wind in my blood-flushed face.

How fast?

The question brings an abrupt end to my reveries, for while I can glance quickly at the speedometer and read off certain numbers, eighty or ninety, say, the words whipping back in the passing wind, I realize, even as I am speaking, that however familiar they may be, neither the words nor the numbers they represent mean very much to me. Eighty or ninety what? Miles *per* hour? Just *how*, a memorable student named Inglefinger once asked me, do you get them hours to go *into* those miles? And for that matter, what is signified by saying that my speed is ninety miles per *hour* if I have not been traveling *for* an hour?

At the far end of the question *how fast?* a number stands shimmering in the heat haze. The analysis of speed involves the specification of that number and so corresponds to the moment when the haze is dispersed by a cool mountain wind, objects standing out clearly in the desert light. Now, no positive speed whatsoever is associated to a body that like an errant brother-in-law recumbent upon the couch is perpetually at rest. An object in motion is one changing its *position*. It was over *there*, in Winnemuca, say. Now it is *here*, in Barstow. Change in position is, in turn, incoherent without a correlative change in time. The tenses tell the tale: it *was* over there, that body, and now it *is* over here.

I am traveling down the highway, in motion, going fast. I am going somewhere. Time is passing. The calculus proceeds by means of the adjuration that change in position must be *coordinated* with change in time. When understood in its full generality, this is a re-markable, an utterly daring hypothesis, one that precariously invests a great deal in the concept of coordination. It is one thing to say, as any old metaphysician might—as I just *did*—that change in position takes place *in* time. It is entirely another thing to say that where a body is *depends* on when it is there, the coordination between *where* and *when* achieved by means of the mathematical concept of a function. *That* is a radiant affirmation of the calculus, a claim it makes about the world.

I am yet going somewhere on the open highway, and yet going fast; but now the highway sliding straight as a bullet across the desert floor, and the blue mountains in the far distance, and even the Harley's iridescent teardrop gas tank winking in the sunlight, reveal themselves, by means of a connection drawn to a Cartesian coordinate system, to be half of a fantastic diorama, the highway matched to one axis of the system, the luridly winking eye of the origin serving to fix a direction along this axis, me on that motorcycle mimicked cruelly by a point, and time's moving arrow mirrored by the system's other axis, an un-yielding straight line. If the motorcyclist would stamp the desert's hot and dusty floor, saying that *this* is real, the mathematician might re-pair to the cool shade of the Cartesian coordinate system in which the desert finds its dried-out image, saying instead with a sly smile that no, *this* is real, the conflict the more vexing inasmuch as motorcyclist and mathematician seem to be sharing cramped quarters in the same body.

But in one respect, at least, the coordinate system is realer than real: it is *there* that the coordination of time and space, which is an inescapable and so an inexpressible feature of ordinary life, is suf-ficiently refined so as to appear an intellectual object in its own right. An intellectual object? Not any intellectual object, of course, but an intellectual object taking form as a function, taking form, in fact, as the *position* function $P(t)$—the position of something *at* time t. Its arguments are various *times,* and its values various *positions*. To say that $P(1) = 3$ is to say that one unit of time has been gobbled up and is gone for good. I have reached the three-mile mark on the open road.

But the position function does more than register the raw record of my travels: it serves wonderfully to *concentrate* experience. The very essence of a change of place is that something was there and now it is here. This essence the position function reveals and then expresses. As its arguments change *it* returns a change in place. The something you say was there then? The *then* in this assertion is expressed by t, and the *there* by $P(t)$. It is, you now say, here five hours later? The *five hours later* is expressed by $t + 5$, and the *here* by $P(t + 5)$. The something that in real life is busy going places has in the position function retained only its phosphorescent smile. Through the mystery of functional notation, the subject of speed—me, in this case—has been swallowed up in favor of its essence. The harsh desert highway, the roaring motorcycle, and the uninformative speedometer now reveal themselves in a flash to be ephemeral aspects of a deeper mathematical reality.

Move Thus or Else

The things of the calculus thus far in place—the real numbers, the concept of a Cartesian coordinate system, the very idea of a function, the position function—these represent the result of a series of fluid and brilliant *gestures,* and like any achievement based on style they remain vulnerable to the corrosive suspicion that behind that gorgeous and fluent show there is not very much of substance. It is entirely possible, for example, that *whatever* the function by which time and space are related, the thing might be uninformative or incoherent or difficult to define or otherwise infirm. Astrologers, after all, point to a connection between the planets in their huts or houses and various human undertakings—love, business, root canal—and as far as they go who could fault them? A connection? Leave the link unspecified and the claim can neither be discounted nor disproved. The functions that the mathematician invokes he must also describe. It is a matter of honor.

The most lucid and accessible of such descriptions has come about by means of a meditation undertaken for the first time by Galileo Galilei. The time is the early seventeenth century. Gone are the low, overhanging clouds of weather-tortured England. Germany and France lie to the north. We are where the lemon trees bloom and the sun traces a line high in the sky. It is one of the pleasures of Italy that almost nothing that is of human importance is the same here as it is anywhere else, not the language, nor the law, nor the lore, nor the people, nor the food, nor the folkways, nor the music; not the color of the sky, not the shape of the trees, not the way the women raise or lower their eyes, and I mention all this in order to fill in the background with a mosaic of familiar facts.

Now imagine a gorgeous tower, its parapet jewel-encrusted, the dreamy Perugian hills in the background, lacy clouds above. And on this tower an Italian dandy, dressed in silks puffed at the wrists and at his thighs, is fingering a large and lavish rose and rufous stone, a fabulous ruby or garnet, something luscious and lustrous. He dangles an elegant forearm over the parapet, holding the ruby in his upturned palm, and then slowly and with vast sensual deliberation rotates his wrist so that the precious stone, its cut facets catching the golden Tuscan light, slides from his polished palm and winking colored fire slips off into space.

After a few seconds or so, how far has it gone? After a few seconds—*time*. How far has it gone—*space*.

An abrupt return from reverie is now in force. The question at issue is at once provocative (How far *has* it gone?) and natural; unlike some questions of the Talmud (*Nu, Rabbi, could the left hand give the right hand a gift?*) *this* question seems of a piece with other questions that idle curiosity might provoke: How far *are* the stars? and Why *does* water find its own level? and Why *is* man born to suffer and then die?

Yet this simple question also tends to prompt a familiar if embarrassed response, which I volunteer on behalf of all those who have not minded mathematics much: *Who knows?*

The facts of the matter, Galileo discovered, are fantastic. Freely falling objects satisfy a relationship between time and space and *only* between time and space. If the world's great engine of change is in this case responsive to time and to nothing else, this means—*does it not?*—that objects falling in Italy and objects falling in Spanish Harlem fall in the same way, their motion fixed by the universal temporal medium in which all action takes place. This is a remarkable conclusion to have reached, an avowal of universality flamboyantly at odds with our sense that most things and especially the things that matter most are relative.

The distance covered by an object falling freely in the air is proportional to the square of the time it has been traveling. Such is Galileo's law of falling bodies: $D = ct^2$, where c is a parameter and t, of course, denotes the time. The prosaic quality of this description is somewhat at odds with the moist mystery that it reveals. By means of purely mathematical operations on purely mathematical objects—*numbers,* after all—the mathematician is able to say that *out there* this will happen or that will. Experiment indicates that c is 16, and so $D = 16t^2$. The units of measurement, I am supposing, are given in seconds and feet. No time having elapsed, t is 0 and D is 0 as well. After one second, D is 16. The falling stone has gone sixteen feet. After two seconds, D is 64. Sixty-four feet have been consumed in change, the stone, whatever it may have been doing in another life, now moving in accord with Galileo's law. Associations to the law and to legal systems have a kind of eerie aptness here, if only to suggest that the discipline of description is itself a source of protection against anarchy.

In both mathematics and life, distance is a number that is inevitably positive; there is no way in which how far someone has gone can somehow bend back to be less than zero. Distance is for this reason congenitally incapable of indicating direction. It is a number expressing *how far* something has gone, but not *where* it has gone. The position function inscribed on a Cartesian coordinate system, by way of contrast, records direction as well as position and so enters the scene defiantly bisexual. With position in mind, Galileo's law of falling bodies takes form as $P(t) = -16t^2$. But the function must really be heard as well as read: *Take the time,* a stentorian if spectral voice commands, *square it, multiply it by this number c. That tells you, Soldier, where on a Cartesian coordinate system an object happens to be.*

The negative sign in $-16t^2$ indicates that whatever is falling is falling *downward* and so is heading for the lower depths. After falling for one second, an object finds itself at -16 feet, a place somewhere below the origin, a remarkable accomplishment *if* the origin is taken as ground zero, and one requiring that a freely falling object tunnel into Hell. It is for this reason that the equation expressing $P(t)$ includes a term indicating the *height* of an object when dropped: $P(t) = -16t^2 +$ **Height**, or $P(t) = -16t^2 + H$. If H is 100, the falling object begins some one hundred feet *above* the origin on the y-axis; after one second $P(1)$ is 84 feet. And that is the position the object has reached.

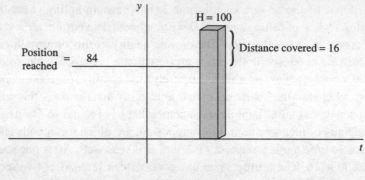

Galileo's tower

As one might expect, the position function occupies itself with the position of a moving point; Galileo's law I expressed originally in terms of the distance the point has covered. It is a difference in formulation marking no distinction in fundamentals. The distance an object has covered may be recovered from the position function as the *difference* in its position at two different times. If the difference is negative, the mathematician recovers a positive number by demanding that the offensive sign be canceled. Such is the *absolute* value of the number, an interesting example of mathematical terminology reverting to the political. Thus $P(1) - P(0)$ is $84 - 100$ or -16 feet. The mathematician's demand is now made, and the magnitude behind -16 liberated as sixteen feet itself. In what follows, this demand is in force so that differences in position and hence distances are always positive.

But all of this belongs yet to the world of words.

Listen

In the background, coming from a vast distance, is the solemn sound of a ticking clock, the measured beats of its great pendulum regular and infinitely soothing. Arising out of nothingness, as all clocks really do, and so from a point conveniently identified with 0, the great clock tolls the numbers: 1, meaning that one second has gone by (or one hour or one year), 2, meaning that two seconds have gone by, 3, meaning that three seconds have gone by, the sounds settling into the troubled human heart and beating as the heart itself beats.

And there suspended before the mind, like an arm of the great god Vishnu, is the symbol $P(t)$.

Let everything you think you know of functions and formulas be canceled and forgotten.

As the clock tolls the seconds, the variable t undergoes a transmogrification, like the transmigration of the soul itself, its crossed arms dissolving and then reforming themselves to designate the numbers: 0, 1, 2, 3, . . . , the graceful ease with which one symbol passes over to another a secret but comprehensible sign of the unity that underlies the world's superficial diversity of appearance.

As the numbers form themselves and then silently vanish into the void, the symbol $P(t)$ changes as well: $P(0)$, $P(1)$, $P(2)$, $P(3)$, . . . , the change in the function felt as a mute throbbing so that what you see you also feel, the rhythms of change entering into the very fabric of sensation.

One second has elapsed; $P(0)$ has gone to $P(1)$.

And there from your vantage point upon a humped and rounded hill, you see that red ruby drop, silent in the sunlight, falling from the tower's top, $P(1)$ revealing its mysterious value, $P(1) = -16t^2 + H$, the symbols no longer functioning as the mathematician's drab detritus, but as imperious and commanding signs controlling from some inaccessible spot the very organization of the world, $-16(1)^2 + H$ telling you in the urgent hot whisper of revelation that in that perishable second the stone has slipped from one hundred to eighty-four feet. In the next, it will have slipped further still, $P(2)$, solemn as the sun when it appears, signifying that in that second the stone has slipped from eighty-four to thirty-six feet, the function acquiring a novel incarnation as a great controlling engine of change, a master of the descending world, the stone and those symbols coordinated somehow, brought into unfathomable alignment, so that what takes place in the dancing air—the jewel-encrusted tower, the Tuscan skies, the Italian dandy, and that heavy falling lustrous stone—are, all of them, under the control of a set of symbols, murmured incantations.

The perishable seconds perish; the symbols abide forever.

And Another Thing

Galileo's law extends its dominion over the earth; its authority carries backward into the past, and it carries forward into the future. And the law goes all the way down. Whatever the time, whether measured in hours, minutes, seconds, picoseconds, or parts of picoseconds, and so on down to the point where the differences between one time and another are like bat squeaks imperceptible to human observers, the distance covered by a falling object remains proportional to the square of the time it has traveled.

To the *square* of time, and not some other function. The function $P(t)$ is a simple polynomial, the result of a conjugal visit between the power function t^2 and an amiable constant. It is nonetheless an object capable of drawing a *specific* connection between time and space, and capable thus of enforcing a stern numerical discipline upon what have been fairly general concepts. It is this enviable specificity that makes mathematics something other than a form of magic. An object falling for three seconds will have traveled precisely one hundred and forty-four feet. The relationship between time and distance is *not* rough and ready. Clocks in the real world may be off by a moment, and measurement of distance always involves some small error, but the *conceptual* connection between time and space that Galileo's law affirms is perfect; it is complete; it is irrefragable. Let the real world of misery and measurement exist in its habitual unchaste state where nothing is ever what it ought to be; the dominion of the law extends over a realm in which relationships are purged of their impurities in an annealing fire. The law marks an ideal. The real world had better get itself together and measure up.

This is a first demonstration that the mathematician's bold insistence that, yes, *his* scheme of classification, one that so far has involved the elementary functions and nothing else, this meager and much-mocked and insubstantial tool, despised by generations of students because it is *so* boring, this plain and prosaic instrument, *is* adequate to the description of an aspect of experience.

A final question, the last for now. Why, the reader may wonder, is it that every falling object chooses to obey *this* law and not some other?

Why indeed?

The Real World's Master

Galileo's law is an achievement in physics, a way of subordinating aspects of experience to mathematical formalism; were the law not known, objects would continue to fall and fall in just the same way, but with the law entered into consciousness, the world of falling things appears to respond to a symbolic taskmaster. *Move thus or else.* Still, the subject at hand is speed. Galileo's law brings about, it reveals, a connection between distance and time. Functions have entered the discussion as a way to represent concepts correlative to speed—change in position and change in time. Speed as a concept has had one aspect of its identity firmly fixed. It is a number. But speed itself has not been specified.

And yet like colored neon lights that may be seen zooming around an illuminated billboard, these far-flung considerations ultimately do cohere to reveal a picture. In the ordinary sense of things, how fast an object is going reflects the *distance* it has covered with respect to the *time* in which it has done the covering. This makes for a vivid if commonplace formula: **Speed = Distance/Time**, or $S = D/T$, to make the notation more compact. Distance is already a concept fixed in the formaldehyde of formalism: *Distance is difference in position,* an aphorism governing social as well as mathematical relationships. And this makes for a way of describing speed entirely in familiar terms:

$$\text{SPEED} = \frac{P(t_2) - P(t_1)}{t_2 - t_1}.$$

Subscripts play a bookkeeping role, keeping the times distinct; t_2 may be read as *the second time,* and t_1 as *the first.* The assumption throughout is that $P(t_2) - P(t_1)$ is positive.

But $P(t_2)$ and $P(t_1)$ are by Galileo's law correlated with

$-16t_2^2 + 100$ and $-16t_1^2 + 100$, respectively, and so like a flower the formula just given unfolds once again:

$$\text{SPEED} = \frac{P(t_2) - P(t_1)}{t_2 - t_1} = \frac{\left(-16t_2^2 + 100\right) - \left(-16t_1^2 + 100\right)}{t_2 - t_1}.$$

Does the formula have a concrete instantiation, an application connected to a physical example? It does, indeed. For speed to accumulate, time must pass. Say, then, that $t_2 = 1$ and $t_1 = 0$. The question I have been insinuating into the discussion at every opportunity, through florid prose and fair—say, fellah, how *fast* is that red and rufous ruby falling?—now admits of a precise, a number-driven answer. That ruby has been dropping at the rate of sixteen feet per second.

Wait, don't go. Here is how the symbols should be read, the answer found.

The distance covered is the difference between positions and so the difference between $P(t_2)$ and $P(t_1)$.

$P(t_2)$ is equal to $-16t_2^2 + 100$ because that is how the position function $P(t)$ has been *defined* by Galileo.

So far, so good? I mean, say something if you don't understand.

But the time at t_2 is 1. Think of a great clock, set to ticking the seconds. At t_2 it has tick-tocked just once.

And so $-16t_2^2 + 100$ is 84 feet: $-16 \times 1^2 + 100$. And that is the *position* of the stone after one second.

You see, it's all really very simple.

But equally, at $t_1 = 0$, the stone has not fallen at all. It is just sitting in the dandy's palm. The great clock has not yet commenced its labors.

And so $-16t_2^2 + 100$ is $-16 \times 0^2 + 100$, which is 100 feet, the position of the stone *before* it is dropped.

Please, let me finish on this, please.

The distance covered by the stone is the *difference* between its positions: $84 - 100$, which is -16, but because of that thing about distance being always positive, I'm going to wipe off the negative sign.

Sure I can do that. It's all right, trust me. I'm a mathematician.

So the distance the stone has covered is sixteen feet, and it has covered that distance in one second. Its speed is—

I know you've got to go. I'm almost done.

Its speed during the first second of its fall is sixteen feet per second.

You see, the symbols say just what they are supposed to say.

For all that, the discussion must now return to the infirmities whence it arose. Infirmities? There is Inglefinger's problem of seeing just how time can be shoehorned into distance. I may as well confess that I introduced *that* as a rhetorical diversion. Time cannot be shoehorned into distance, or shoehorned into anything at all. Division is a concept applicable to the numbers and only to the numbers. In saying that a stone is falling at the rate of sixteen feet per second, the relevant ratio is between the *numbers* 16 and 1; the units are just there for the ride.

Nonetheless, the concept of speed that has emerged from these considerations seems strangely splayed still, depending as it does on the time things get started and the time things end. There is in this appeal to the temporal extremities of action something unnatural, perhaps forced. In the development of the calculus, this careless observation marks the spot where a sense of intellectual anxiety commences, a deep breath drawn.

The Real World's Servant

Galileo was born in 1564 and died in 1642, six years before Descartes. He is now among the immortals—the men who made modern physics; but at his death, he kept company in the hereafter only with Copernicus and Kepler, Descartes still freezing his behind off in Stockholm, Newton busy being born, and the rest of the immortals patiently waiting their turn by the cosmic wheel.

It is in his dialogue *The Two Chief Systems*, written in 1632, that Galileo refined and transmitted to an only half-believing world the Copernican doctrine that the earth revolves about the sun. He was, for his troubles, denounced by the Inquisition, and, in a moment now famous as myth, forced on his knees to recant; but as he rose, humiliated by the proceedings and no doubt frightened, he was heard to murmur *eppur si muove—yet it moves;* and so aligned himself forever with the defense of objectivity against superstition.

For all his greatness as a physicist, Galileo's mathematical abilities were incomplete, his intuition often misleading, this despite his conviction that "nature's great book is written in mathematical symbols." He was an indifferent technician, his powerful imagination undisciplined. Fascinated by the infinite, Galileo imagined that some novel form of infinity lay between the finite and the infinite numbers. Not so. He grasped quite clearly that the even numbers 2, 4, 6, 8, . . . , and the natural numbers 1, 2, 3, 4, . . . , could be placed into correspondence, one natural number for every even number, and so saw clearly that the infinite *parts* of an infinite collection might be as numerous as the collection itself. This is not a paradox but simply a property of infinite collections. Galileo could not imagine that infinite magnitudes might differ in size, as, in fact, they do, the real numbers forming an essentially *larger* set than the natural numbers or even the rational numbers. This idea required Georg Cantor's mathematical genius for its expression, Galileo missing entirely the low hole in the wall that leads to modern set theory. There is something endearing about the quality of Galileo's mathematical pottering. Gifted enough to sense, he remained unable to see, and so, unlike Descartes, his contemporary, he passes into mathematical history as a figure on the margins. Like Einstein, Galileo was able to reach fundamental physical conclusions *without* the assistance of mathematical genius.

It is the aged Galileo, the Galileo of the confrontation with the Inquisition, who matters most. The idea that the universe possesses an intrinsic, distinct, and objective character is known in philosophy as *realism* and in ordinary life as common sense. It is an idea that in the late twentieth century has come under attack. Literary critics see texts receding behind texts, the objective world vanishing in a whorl of words. Philosophers of science? They busy themselves by calling attention to the sheer conventionality of scientific thought, paradigms piling up behind paradigms. There is no *progress* in the passage from paradigm to paradigm, only the rouged lips of fashion. Analytic philosophers urge upon one another lurid and revolting varieties of *irrealism,* the world, on their view, flickering into existence inconclusively, as much *made* as discovered, as much *fabricated* as found.

And yet there *he* is, Galileo, the gentle servant of the real world, trudging forward patiently up through the hours, days, and years, insisting always in his own slow, stubborn, and defiant way that regardless of what authority might affirm or fashion dictate, there *it* is, out there, objectively distinct, indifferent to human intervention, the universe, the world itself, the commanding thing that we are meant to know.

chapter 13

Speed, Strange Speed

As the calculus commences, the mathematician stands within the confines of a Cartesian coordinate system, snake stave in hand, his head turned at a ninety-degree angle so that he may contemplate Ra, the hook-nosed and Unchanging One, and *we* on the outside of the frieze feel, as things begin, a rumble of impatience: *Don't just stand there, fellah, do something.*

Do something? *Do what?* The man is a mathematician. Let him *do* something mathematical. He has already made position a function of time, the severe formality of the frieze acquiring in this way a hint of living life. With the present invocation of speed, the mathematician does something more. Functions may flicker in the unyielding sky, but ask how fast things are changing in their places, and the Cartesian world in all its steadfast gorgeousness undergoes a dramatic transformation. Directly, the crude, vulgar, but violently *alive* sound of a motorcycle exhaust is heard. *Look at that sucker take off!*

The idea that speed should be defined as a ratio insinuates a familiar concept into the calculus—a *ratio,* as in one thing compared to another (distance and time), a ratio, as in one number divided by another (the difference in position by the difference in time). But however defined, the concept that results is too clumsy to be of direct use in the calculus. Thick-necked Roman troops tramping along the Appian Way bragged, no doubt, about covering twenty miles a day on their forced marches. I have been going faster, but what they meant is what I mean when I recount to my father-in-law that, *you know, Bob, I was doing seventy the whole time I was on the interstate.* What we have in mind, those legionnaires and me, is a concept contingent upon *two* distinct times (*when* I started and *when* I finished) and thus upon *two* distinct positions (*where* I started and *where* I finished), and so a concept representing an *average* of sorts. In saying that I did seventy on the interstate, I make my claim only with respect to the distance I covered *in* an hour, the question how fast I was traveling at any particular time left indeterminate.

This last remark prompts a worm of doubt to wriggle. I might have covered seventy miles in one hour by doing a forbidden eighty-five during the first half-hour I was traveling and a legal fifty-five for the half-hour that followed; for that matter I *might* have covered those seventy miles by flying at four hundred and twenty miles an hour for ten minutes (in a jet helicopter, say), remaining placidly at rest for the fifty minutes thereafter, the helicopter parked by the side of the road, its rotors spinning slowly. That ecstatic ten-minute burst notwithstanding, travel by helicopter yields the *same* average speed as my normal guilty surge and shuffle along the freeway, the two quite different ways of moving myself along resolving themselves into a common velocity.

And needless to say, there *is* something odd and even disturbing about a concept in which the speed of an object is not distributed to each of the times in which the object is moving. The pale-eyed highway patrolman who has just stopped me for doing eighty-five miles an hour on the Bayshore Freeway and is now standing by my elbow, his leather gun belt creaking in the autumn air, *he* remains curiously unimpressed by my insistence that the eighty-five registering on his speedometer is merely an anomalous and high-spirited spike, my anticipated *average* speed, when slow traffic is taken into consideration, a safe and sedate fifty-five miles an hour. He looks at me with amused contempt as my explanation sputters and then stops.

"I'm not interested in the distance you figure you were going to cover in an hour, sir," he says, turning my license over and then frowning because in defiance of the plainly written instructions on its bottom margin, I have encased the thing in Lucite. "I pull you over you were doing eighty-five."

"Don't argue with him," hisses my companion in crime, her injunction echoed by generations of mathematicians, who, I imagine, have been watching this scene with malicious enjoyment.

The highway patrolman's insistence that all matters between us be made relative to the particular moment—*the very instant of time*—at which he happened to mark my speed is a brusque reminder from the real world that quite often an object's average speed is too coarse a concept to deal with the day-to-day contingencies of even so resolutely literal a subject as law enforcement. The patrolman's abjuration is echoed by that great competent chorus in which men habitually deal with speed, our airline pilot remarking that because *we're picking up a little tailwind there, folks, we're doing just over six hundred miles an hour,* meaning, of course, that as we pass over twilight-flooded, field-checkered Kansas, Grain Ball City visible as a proud preposterous dot, our speed *right now* is a little over six hundred miles an hour, the same voice, heard at a Grand Prix racetrack, reporting with the same maddening aplomb that *going into that last curve, folks,* a rendezvous with a bathtub full of beautiful blondes only moments away, a driver named Mrcs or Snrxs, these people for some reason having dropped the vowels from their last name, *seemed to top out at two hundred and eighty miles an hour,* the impression of speed made vivid, indeed indelible, because for Mrcs or Snrxs the journey into the last curve has ended abruptly at the retaining wall, Mrcs or Snrxs tottering out of his shattered car and patting effetely at his flaming suit, the point in all this that at the very moment he hit the wall, this lughead was traveling at two hundred and eighty miles an hour. *At that very moment.*

The precise pairing of speed and time yields an object's *instanta-

neous velocity, and with its introduction an entirely new concept undergoes the first primordial throb of life. Whatever its roots in the real world, instantaneous velocity arises from something more than the quotidian desire to compress speed so that it settles on a point. It is a consequence of concepts already in place and intellectual choices already made. *Time and space, change and position, distance and speed*—these are related ideas, part of a great wheel. They are concepts that express relationships between real numbers. But the relationships between real numbers sanctioned by the calculus are those that may be expressed by a function, an intellectual instrument that broods over the mathematical world. A concept that does not admit expression in these terms ceases to exist. Insofar as speed finds itself cast upon the great wheel, its very expression within the calculus is from the first moment foreseen as a function: *some speed must be assigned to each moment.*

Leibnitz and Newton alike understood the subordination of speed, how it must be expressed in the context of tightly related concepts, but writing on speed Newton is often difficult to follow, his thoughts not so much insufficiently clear—I am, after all, talking of Newton—as insufficiently organized: he introduces fluents and fluxions, he writes of sequences and limits, he gives every indication of having seen more than he could reveal, and in the end, more often than not, he settles physical matters by means of his superb intuition. It is in Leibnitz that we can sense the light gather and see a concept coming into creation.

Leibnitz Meditates in His Room at Night

Sitting in his study in Hanover, the brown-walled room close from an elegant and very efficient porcelain stove, he must have *first* said to himself, rehearsing the obvious points as he adjusted his plump posterior into his ornate wooden chair, vellum sheets spread before him on a leather-covered desk, that speed involves a change in place and a change in time—and here Leibnitz moves his fleshy right hand in front of his face in an arc from left to right, palm inward, his hand pivoting on his elbow, as if to remind himself that as it has flesh and blood, some tangible thing in the real world *has* speed.

It is well past midnight, the deep quiet of the provincial city seeping into the sepia study. A clock chimes the quarter-hour. A feeling of intellectual fullness comes over Leibnitz, a calm sense of his own mind's spaciousness. The arc that Leibnitz has been tracing with his hand, he contracts so that only half the distance is covered in the space before his face.

He leans back into his chair and holds before his attention, like two raddled courtiers, those two abstractions: Change in Place and Change in Time. The sounds of the faraway clock are routine and reassuring, the regular division of the temporal continuum, one point following the next. He holds his hand steady, the fingers relaxed. In moments such as this, Leibnitz has a tendency to address himself. *What,* he asks, is the speed *now*? His thoughts move slowly through the great room of his mental powers. He realizes, he has *always* realized, the discordance between speed, time, and distance. Beneath the appearance of contradiction in these concepts, he senses some scheme of reconciliation.

The steady rhythmic ticking of the clock prompts Leibnitz to imagine the sounds suspended in space: he seems to see what he hears, each gentle tick breaking open before his eyes like a small multicolored burst. He attends to one such burst, the sound (or the sight) marking the moment as it arises and then perishes, and with the thumb and forefinger of his left hand, he measures the distance between that sound, which has already vanished, and the one to come. And for a moment, he sits in that strange position, his right hand describing a gentle arc, the fingers of his left hand, measuring the time, drawing close to one another.

Whatever the distance, he thinks, half forming the words, thinking almost in images, it is yet *some* distance that my hand has moved, whatever the time, it is yet *some* time that my fingers have measured. Leibnitz looks closely at his own well-made hand, the left one resting on the desk's edge, a manly network of veins running from the top of his wrist forward to his fingers. *Aber sicher.* To be sure. And yet what if the distance between times were to become small?

How small? Leibnitz asks himself.

The clock chimes, and as it does, Leibnitz taps the table-top with his thumb. *This is now,* he says.

With great delicacy, Leibnitz allows his forefinger to approach his thumb. *Very small*. He can sense in the skin between his fingers the contracted space between his flesh. *Infinitely small*.

He sits like that for a heavy, pregnant moment.

Meaning what? The distance between times is infinitely small. *Yes, infinitely* small, but still, it is not nothing, that distance. And if that distance is not nothing, *it must be represented by a number*. Not a genuine number, he thinks, adjusting his thoughts to account for his own audacity, but a useful fiction, an imaginary object. And it must be *smaller* than any other number.

Is it 0, then? No, that would make nonsense of everything, division by zero an invitation to the black hole of meaninglessness. *Smaller than any other number, but yet greater than zero*. Interesting. An infinitesimally small number measuring an infinitesimally small distance in time.

Leibnitz eases his legs underneath the table. *Infinitesimally small numbers?* His eyes are heavy. The hour is late. There were times when he would sit like this throughout the night. *And why not, after all?* For a moment, he lifts his head as if to address an audience. He maintains contacts throughout Europe with over two hundred correspondents. At night, he imagines that he is conversing with them. At the one end of things, the natural numbers 1, 2, 3, 4, 5, . . . , form an infinitely large progression. *Why not a compression of numbers so intense as to be the inverse of the infinitely large?* It is a large step to take, but beyond this step, he can see the rest of the calculus unfold; in such situations, Leibnitz thinks, switching almost automatically from German to French, *il n'y a que le premier pas qui coûte*—it is only the *first* step that counts.

Leibnitz raises his hand again, traces an arc from right to left and then from left to right and holds his hand steady midway through the arc.

Taking up his heavy gold pen and smoothing out the gray vellum sheet with the linen cuff of his sleeve, Leibnitz writes the symbol *t*. This stands for the time *now*, he says emphatically to himself. Then he writes the symbols *dt*. Now let *dt*—yes, let *that* be the *infinitesimal* difference between the time now and the next moment in time.

With the unselfconscious mental habits of a man used to abstractions, Leibnitz allows the soft brown of his study—the superb walnut desk, with its rich inlay of ivory, the drapery-covered walls broken by formal portraits of his patrons, the yellow and white porcelain stove—to dissolve and imagines instead the axes of a coordinate system stretching in front of him, the origin, he realizes with a soft smile, aligned precisely with his large nose.

Dipping the quill of his pen into a crystal inkwell, he then writes the symbols dy on the vellum sheet, the pen making a soft scratching sound. *Now then,* he says sternly to himself, if dt is an infinitesimal difference in time, I must say that dy is some infinitesimal distance in position—how far my hand has traveled in an infinitesimal amount of time. Leibnitz moves his hand in its by now familiar arc, imagining that it is representing changes on the y-axis of the coordinate system, but now he tries to make the movement of his hand constrained; he notices as he does a slight tremulousness along his fingers.

He sits there, a large portly man in a brown room, staring at a coordinate system, the coordinates fading finally into the warm sepia of the room itself, long shadows lit by a single candle.

Then Leibnitz writes the symbols dy/dt on his vellum sheet. He has always believed in the power of symbols. "Good symbolism is one of the greatest aids to the human mind," he says quite aloud, quoting himself, addressing his congregation of European correspondents. He fixes his attention on the symbols dy/dt , allowing them to form and reform, the letters separating from one another, elongating themselves. Then he draws a heavy black line underneath the symbols. The ratio of a change in place during an *infinitesimal* change in time, *this,* he says, is the speed an object is traveling at any given moment t. The great wooden grandfather clock in the downstairs hallway tolls the hour, and as the last mournful chime reverberates, Leibnitz imagines, he actually *sees,* time moving forward from that chime by an infinitesimal amount, a tiny dimpled hiccup in its everlasting forward flow.

He looks down to consider again what he has written—the letters dy/dt .

The first of the great symbolic forms of the calculus has taken shape on those vellum sheets sometime late at night, sometime late in the seventeenth century.

The Future Rebukes the Past

There is to this daring idea, and especially to the notation in which it is expressed, a strange and moving power. It succeeds, this definition of instantaneous velocity, in its first purpose, which is to assign a speed to every moment of time: it accomplishes the subordination of a wayward concept. It succeeds as well in showing that average and instantaneous speeds are comparable concepts, both expressed as ratios, both revealed by the same mathematical operation. And above all, it succeeds in showing that something so simple as one's hand moving gently in the warm air requires for its analysis things that are infinitely small and so serves to remind us that we are like mariners on a strange blue sea.

It remains only to be observed that the idea of an infinitely small number, and so the very concept of an infinitesimal, is somehow absurd.[1] And in this, the future turns to rebuke the past.

You must imagine them, standing there together on the stage of history, hands clasped behind their backs, looking off into the middle distance, the mathematicians of the seventeenth and eighteenth centuries, Leibnitz and Newton, of course, but the others as well: d'Alembert, L'Hospital, and Lagrange, dressed in lace and ruffles, perfumed, wigged, and powdered. They have made the calculus, and what they have made works brilliantly. They are simply unable to explain what they have done. A voice calls from the audience:

What precisely are such quantities as infinitesimal distance, infinitesimal time?

"They are numbers, to be sure."

Are you certain?

"Very much like numbers."

"Ideal elements, actually."

"Or rather like numbers but perhaps not quite like any of the other numbers. Fictions."

Fictions?

---

[1] Demonstrably so, in view of the appendix.

"Actually, not quite fictions. Perhaps it would be best to describe them as useful fictions."

How can a fiction be useful?

"Well, they have been useful, haven't they? So it follows that they must be useful."

You are serious?

"Quite serious. But, of course, we are mathematicians."

Can you describe these infinitesimals, tell us anything more about them?

"They behave, really, as if they were numbers, but, of course, they are smaller than any of the other numbers. Otherwise, they would not be infinitesimal."

"Quite right. Small, very small."

Zero, then?

"I say, not quite zero, larger than that."

"But not much larger."

Could you be more specific?

And the remarkable thing is that in answer to this last question this collection of brilliant men could do no better than to thrust their hands into their pockets, allow their cheeks to be suffused with blood, and stare somberly upward at the ceiling. Writing in 1734, Bishop Berkeley wasted no time in attacking the very idea of infinitesimals. If they *were greater than zero,* the definition of instantaneous velocity would not define anything instantaneous, and if they were *zero,* the definition of instantaneous velocity would not define anything like speed. Berkeley was himself a philosopher, best known for his view that objects exist only insofar as they are perceived and thus not a man unfamiliar with absurdity; his arguments were acknowledged, but after a momentary guilty pause to see if anyone would pay them attention, the great mathematicians of Europe simply accepted the infinitesimal calculus, coming to regard infinitesimals with the sense that in this respect they seemed to have gotten away with it all.

Looking backward, *we* can see that Berkeley was entirely correct. There are no infinitely large or infinitely small numbers. And this injunction the development of the real number system itself enforces, the twentieth-century future coming to rebuke the eighteenth-century past. Whatever the real number *r*, a theorem sings out, there is a natural number n such that r is less than n. But then the theorem immediately implies that for any real number r there is a natural number m such that $1/m$ is *less* than r. This rules out of court infinitely small numbers, just as intuition demands, and with it an analysis of instantaneous speed based on the infinitesimally small.

The road toward a definition of instantaneous speed, which had that evening in Hanover seemed straight, now looks crooked as the very jaws of Hell.

Nothing Is
Infinitely Great
or Infinitely Small

Read this. *No, no,* don't try and skip ahead to the good stuff. This *is* the good stuff. The essential idea behind the banishment of infinitesimals is very simple. Whatever the real number r, there is a natural number n that is bigger than r. No real number is biggest; and, of course, none smallest. Such is Archimedes' theorem. (The attribution is somewhat suspicious but the name has stuck.) Now instead of talking of the numbers r and n, I might convey the effect of the theorem by saying simply that the natural numbers $1, 2, 3, 4, \ldots,$ are *unbounded* amidst all of the numbers.

And here a drama of delicate distinctions begins. *Unbounded* meaning there is no largest natural number? Surely *that* is obviously true. Whatever candidate n might be proffered, $n + 1$ is larger yet. The requisite sense of *unbounded* lies elsewhere.

Imagine numbers collected together in a set—K, say. K is *bounded from above* if there is a number x that is greater than or equal to any number in K. This is a definition. In symbols: $x \geq a$ for each and every a in K. Such an x is an *upper bound* for K. If K comprises the elements $\{1, 2, 84, 34, 3, 17\}$, then 84 is an upper bound for K, but so is every number larger than 84.

A second definition follows. A number x is a *least upper bound* for K if x is least among the upper bounds. Symbols require a double declaration: $x \geq a$ for each and every a in K: and if y is an upper bound for K, then $x \leq y$. The number 134 is an upper bound for K, and so is 193, but 84—*that* is its least upper bound.

Dedekind's axiom brings the real numbers into existence; but like so many important foundational assertions in mathematics, it has an equivalent formulation, an elegant double. Any set of numbers that has an upper bound, Dedekind's double affirms, has a least upper bound. Such is the Axiom of the Least Upper Bound. And, indeed, a recollection of Dedekind's axiom itself suffices to reveal the strong family resemblance between these two formulations, a Dedekind cut bringing into existence a real number that functions precisely as the least upper bound for the set below the cut.

These distinctions in place, the meaning of *unbounded* may now be made plain. The set of natural numbers, Archimedes' theorem says, has no upper bound amidst *all* the numbers, the real numbers included. For suppose that it did. Then by Dedekind's double, the natural numbers would have a least upper bound—a, say. By definition, $a \geq n$ for any natural number, but then $a \geq n + 1$ since $n + 1$ is a natural number if n is. Subtract 1 from both sides of this last inequality. It follows that $a - 1 \geq n$ for every natural number.

And from this it follows that $a - 1$ is *also* an upper bound for the natural numbers. But look now at this: $a - 1$ is less than a, but a was said to be a *least* upper bound for the natural numbers.

With the revelation of this contradiction, infinitely large numbers vanish into the void. An obvious adjustment in the argument gets rid of infinitely small numbers. And this is true wherever an Archimedean theorem holds sway.[2]

---

[2] In the non-Archimedean fields it doesn't. The development of such fields by the logician Abraham Robinson in the twentieth century has made possible the development of the calculus entirely along the lines anticipated by Leibnitz. But at a very great price in plausibility.

chapter 14

Paris
Days

As the nineteenth century began, the French mathematician Augustin-Louis Cauchy was but eleven years old, and this odd curiously adhesive little fact serves to circumscribe the wheel of mathematical change within the great turning wheel of revolution, war, and social disorder. Fifteen years later, the waves of Napoleonic conquest having swept over the Continent and then receded, the Congress of Vienna created throughout Europe a system of stable if repressive regimes; in the long moment between the end of one era and the commencement of another, Augustin-Louis Cauchy appears on the streets of Paris as one of the men who fully meet their time, the first of the modern mathematicians, connected (in my mind, anyway) by a curve in the current of history to Peter Abelard, who in the twelfth century gave lectures on logic from the hilltops of Paris in moments snatched from kissing Héloïse, the two men cagey and careful on matters of definition and detail, their caginess curious because so few figures in the history of thought have fully appreciated the queer power inherent in the correct, the stable, formulation of an idea.

A graduate of the school of bridges *(Ponts et Chaussées),* one of the remarkable educational institutions created by Napoleon, Cauchy seems to have been sustained throughout his professional career by the force of a barely controlled, a quivering, intellectual passion. His was a great *organizing* intelligence. There is in everything that Cauchy wrote a forward-looking thrust, a comprehensiveness, a sense of *professional* classification that is inescapably modern.

Cauchy wrote more than eight hundred papers, and there are in contemporary mathematics traces of Cauchy almost everywhere, theorems, definitions, solid and accurate proofs, the great textbooks he wrote in the 1820s—the *Cours d'analyse,* especially—having set the standard for mathematical education for the next one hundred years. The form they gave to the calculus is the form it yet retains, so that in a queer reversal of time's arrow, it is often impossible to consult an original paper by Cauchy without wondering somehow whether the man had access to the modern textbooks in which his ideas now lie ceremonially entombed.

The Cauchy of the calculus is the Cauchy who turned his frail back against the infinitesimals of the eighteenth century; it is the Cauchy of the great definitions, the definition of a limit essentially *his* creation and as much of a miracle as those fantastic Swiss clocks of the period in which hundreds of gleaming cogs are made to celebrate not only the time and date but the phases of the moon. The universe, modern science often suggests, is stranger than it seems. Things are said to begin with a bang; space and time are curved like a bright bow; and when after endless time has passed the great stars collapse, they leave behind black holes into which matter and light tumble and then vanish. How much of this is true, how much enchanting fantasy, I do not know; but quite before the creation of modern cosmology, the calculus serves to demonstrate with an eerie aptness the extent to which ordinary concepts are not ordinary at all. Simple speed seems a concept on the margins of the infinite, and yet the strangest thing of all, stranger by far than those black holes in space, is the fact that the cat's cradle of words that Cauchy offered the world *is* sufficient to purge speed of its paradoxes.

The direct, unforced, natural, intuitive, and persuasive definition of instantaneous velocity—the one that Leibnitz offered to his audience in the night—*that* history has canceled and rebuked, the theory of infinitely small or infinitely large numbers replaced thereafter by the theory of a limit, the moment in time when the new theory supplants the old as redolent and as resonant as that other moment in the history of science when after two thousand years the great and complex Ptolemaic theory of the heavens was replaced by the modern, Copernican and Galilean theory, the heavens at once regaining their lustrous simplicity. But it is an artifice, this new concept, *not* direct, unforced, natural, intuitive, or persuasive, and as an artifice the source of intellectual anxiety, otherwise capable men and women remembering their undergraduate or high-school experiences with a moan of misery. *Limits?* The moan darkens.

But now a professional secret must be imparted. The *concept* of a limit is simple. It is the *definition* that is complex. The concept involves nothing more obscure than the idea of getting closer and closer to something. It suggests the attempt by one human being to approach another: and the inexpungeable thing in love as in mathematics is that however the distance decreases, it often remains what it always was, which is to say, hopelessly poignant because hopelessly infinite.

An Artifice of the Infinite

A *sequence* is a set of numbers in a particular order. The numbers 1, 1/2, 1/3, 1/4, 1/5, . . . , constitute a sequence S_n, one that goes on forever as its terms get smaller and smaller. S is the name of the sequence, and n denotes its nth term, S_3 marking 1/3, S_4, 1/4, and S_n, 1/n, this last, in effect, a general rule, a kind of recipe, governing the construction of the sequence. At the 10,526th position in the sequence—10,525 terms having swum into sight and 10,525 terms counted—the rule affirms that out in the distance the value of the 10,526th term is 1/10,526.

The even numbers 2, 4, 6, 8, . . . , constitute another sequence R_n; it is one that like S_n goes on forever, with each term obtained from the one that has gone before; the requisite rule asks only that the number 2 be added to the last term in the sequence in order that the next term be formed.

Still, there is an intuitively palpable difference between these sequences, one that is evident when they are exhibited together:

$$R_n = 2, \quad 4, \quad 6, \quad 8, \quad 10, \quad 12, \quad 14, \quad 16, \ldots,$$
$$S_n = 1, \quad 1/2, \quad 1/3, \quad 1/4, \quad 1/5, \quad 1/6, \quad 1/7, \quad 1/8, \ldots$$

The first ascends monotonously, pair by pair, a slowly growing arithmetic progression, the animals trooping aboard the Ark; the second descends delicately, moving downward from 1, sharply focused; but beyond the differences in their direction, there is the obvious fact that the second sequence is approaching a *goal* or *target* or fixed *point,* for as the numbers continue, the terms of the sequence come closer and closer to a limit at the number 0, the sequence as a whole tapering downward toward 0 like a driven spike.

These distinctions are prompted by real mathematical facts. And the facts in turn contain almost everything that is relevant to the idea of a limit. The first sequence has *no* limit; it keeps lumbering on. The second sequence has a limit at 0, which it *approaches* or toward which it *converges*.

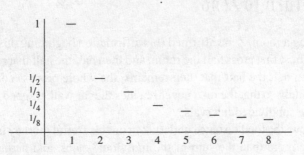

The sequence $S_n = \frac{1}{n}$ approaching a limit at 0

The facts conveyed by the picture mathematicians symbolize in their own limpid vernacular:

$$\lim_{n \to \infty} \frac{1}{n} = 0.$$

The limit of the sequence $1/n$, the symbols say, as n gets larger and larger, is the number 0. The very elegant $n \to \infty$ is read *as n approaches infinity,* the arrow and the flopped letter eight signifying that n is getting larger and larger and thus large without end. S_n *has* a limit at 0, if as n approaches infinity, S_n approaches 0.

The example afforded by $1/n$ suggests the meaning behind the metaphor of a mathematical approach. It is just this: that as the sequence is extended, the differences between S_n and 0 become smaller and smaller. The difference between 1/4 and 0 is 1/4. The difference between 1/8 and 0 is 1/8, and the difference between 1/10,526 and 0 is 1/10,526. The sequence is getting longer; *differences are getting smaller.* In this way, a mathematical operation (taking differences) comes to supplant a vivid mathematical metaphor (approaching a limit). The intellectual movement encouraged by the definition is one of expansion (the sequence being extended) and contraction (the differences becoming smaller). Its basic rhythm is one of breathing in and breathing out.

A Return to Zeno

Crossing a room, Zeno affirmed (his affirmation beginning this book), a man must first cross half the room, and then half the half that remains, and then half the half that then remains, the whole process continued indefinitely so that the man never reaches the far wall, trapped forever in those subdivided intervals.

Now, the total distance undertaken on this fretful journey is surely nothing more than the sum of the individual steps, and perhaps *that* can be represented mathematically by a queer kind of object, an *infinite* sum or series:

$$\frac{1}{2} + \frac{1}{4} + \frac{1}{8} + \frac{1}{16} + \ldots + \frac{1}{2^n} + \ldots,$$

one whose terms correspond in the obvious way to the successive sub-division of the distance between one end of the room and the other.

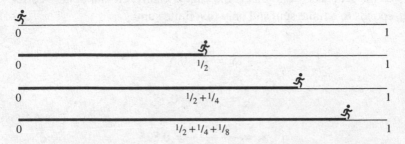

Crossing a room with scant success

Evidently this sum, based as it is on an infinite number of terms, must itself be infinite. Hence Zeno's hasty but very natural conclusion that insofar as the distance crossed is infinite, the crossing never ends.

It is thus that matters stood for more than two thousand years, Zeno of the paradoxes present in a thousand chalky college classrooms, but not where matters now stand. Addition is an operation, experience suggests, that makes sense only when finitely many numbers are added, the imagination straining at the very idea of totaling an infinite column of numbers. But beyond the world divulged by experience there is another world, one divulged by definitions. It is the concept of a limit that makes possible the concept of infinite addition, one concept coming to generate another in that extraordinary way in which ideas once introduced take on a vibrant and then a pullulating life of their own.

The essential idea is simple and elegant, inventive and ingenious. Infinite addition the mathematician proposes to *define* in terms of sequences and their limits. The drama that results has some of the thrill of a high-wire act performed without a net. There on the page is Zeno's infinite sum or series, standing as a perpetual rebuke to the possibility of motion. The mathematician is determined to *compel* those infinitely many numbers to reach a finite sum. There follows now a blaze of misdirection. Infinite addition proceeds *via* the introduction of a sequence S_n, one whose terms are partial but thoroughly finite sums. Thoroughly *finite,* note, that is to say, thoroughly *ordinary.* The first term in the sequence corresponds to the first term of the series, the second term, to the *sum* of the first and second terms of the series, the third, to the sum of the first, second, and third terms of the series, so that *each* term of the sequence—and this is the cagey, canny, crucial point—corresponds to a finite sum and only to a finite sum:

$$S_1 = \frac{1}{2} = 0.5,$$

$$S_2 = \frac{1}{2} + \frac{1}{4} = 0.75,$$

$$S_3 = \frac{1}{2} + \frac{1}{4} + \frac{1}{8} = 0.875,$$

$$\vdots$$

$$S_{16} = \frac{1}{2} + \frac{1}{4} + \cdots + \frac{1}{2^{16}} \approx 0.99998474.$$

The mathematician now allows this sequence inexorably to advance. As the sequence goes on and on, pale and partial sums appear to be approaching the number 1, with S_{16} less than 1 by only a whisker (the symbol $\approx$ indicating approximate equality). Now an appearance is not yet a mathematical fact. Those partial sums may well barrel on past 1. But suppose not. Say instead that 1 is the end of the line; say, in fact, that

$$\lim_{n \to \infty} S_n = 1.$$

And now the mathematician engages in one of those liquid imaginative leaps that suffices to characterize mathematics as a performing art. The *limit* of the sequence S_n, as n approaches infinity, the mathematician *assigns* to the infinite series as its sum. *Assigns?* Meaning what? *Assigns* meaning *decides that it is so, assigns* meaning *acts,* as in the mathematician acts to create sense where before there were only those sums tramping on and on.

Assumptions having backed up behind the mathematician, this may not suffice to settle Zeno's paradox; it does show that insofar as the paradox rests on the assumption that an infinite sum *must* itself be infinite, the mathematician still has room in which to maneuver, the concept of a limit prying open a panorama, allowing the mathematician for the first time to see unseen and forbidden things and suffusing with a hard, utilitarian light exercises in thought that had long been part myth and part mystery.

And there on the streets of Paris, did Cauchy foresee all that his definition could do? More so than most mathematicians, the man has a way of disappearing behind his masks. Cauchy demonstrated throughout his professional life a deep commitment to the Bourbon house in French politics and remained thus a royalist amidst republicans, his conservatism very much of a piece with his Catholicism.

I never knew, or understood, what an affiliation to the idea of monarchy meant until I lived in Vienna. Once the last living member of the royal Habsburg family—the widow of Charles I, actually—paid a visit to the city. For days, the papers talked reverently of little else. I wandered by the square in front of the Stephanskirche just before she was to arrive. The usual curious and indifferent crowd was there, of course, but toward the center of the square I could see that the spectators were older, dressed impeccably. They stood waiting patiently, the men with their chests puffed out, wearing Austrian hunting hats with feathers, the women frail, like mannikins, and tidy. An honor guard decked out in red and gray Habsburg uniforms was at attention. A drum roll began. And then, inching its way along, a coach drawn by six horses passed through the crowd and came to a halt. At the back of the square, where bright teenagers clustered, red rouge on their cheeks, or wispy beards, there was an aching sense of expectant tenderness, until the coach door opened, and a tiny, wizened, terribly old woman, dressed somberly in black, descended from the coach helped by four enormous footmen, stood erect in the square, and with glittering, glaring eyes looked out at the crowd and for a moment smiled, her ancient lips wrinkling.

A man of competitive piety, notorious for his solemn proselytizing, Cauchy, it was said and often said meanly, would fasten on anyone, regardless of station or rank, and speaking in that low drone characteristic of obsession, outline in great and complicated detail the advantages of the Catholic system of belief. On seeing him approach with that mad muted look in his eyes, friends would brace themselves, for he was relentless; acquaintances would simply flee, his clacking entrance into the Academy of Science sufficient to empty the great hall with its glistening gilt mirrors; yet his anachronistic political ideology and baroque and unyielding religious sentiments were apparently compatible with his scientific principles, and this may suggest, as it does to me, that the common assumption that scientific genius inevitably inclines a man toward agnosticism in his religious or political convictions is little more than a modern myth.

Cauchy spent the greater part of his life in Paris; his work reflects the city and in the end the man became the place. It is hard-edged, bright, various, brilliant, and civilized, the work, the blue northern light sifting through the pages of his endless papers, luminous and cold. And it is thus that I myself encountered Cauchy, walking briskly down the street, not one of the grand boulevards, which are on the right bank, but the somewhat seedy Rue Monge on the left. He is by contemporary standards a slight man, thin throughout his trunk; his elbows are held against his sides and he is walking rapidly, with his head thrust forward, the hard wooden heels of his shoes clacking on the pavement. He is hurrying to a meeting at the Academy; he is forever hurrying and he is forever distracted. He is thinking about an issue in the theory of elliptic functions, and he has already attended in his imagination to the monograph that will express his thoughts. Something perturbing has just occurred to him; he stops by one of those by now old-fashioned green *pissoirs* that can still be seen in Paris and as a workman in blue overalls inside thoroughly eases his bladder, there Augustin-Louis Cauchy, friend to the House of Bourbon and a devout Catholic, for a moment closes his burning eyes.

The Limit
of a Sequence

Informal remarks do not constitute a definition: they serve only to convey an idea. The precise definition of a limit is one of civilization's accomplishments, but there is no evading the fact that the definition makes genuine intellectual demands.

A sequence S_n has a limit at the number L if, as the sequence is extended, its terms get closer and closer to L. So far, so good. There are two mental motions here: the extension of the sequence and its convergence toward a limit. And what makes any definition to come still more rebarbative is the fact that these mental motions are somehow coordinated.

A revised but still a vernacular version of the definition now follows. A sequence S_n converges to L if, by extending the sequence, the distance between S_n and L may *indefinitely* be decreased.

The image is of a boat on the deep blue sea conveyed toward a lighthouse by the motion of the waves, the distance between ship and shore being inexorably sliced away. This may suggest, as it suggested to generations of mathematicians, that convergence hinges only upon some fixed but very, very small distance. It is at this point that the translucent ghost of infinitesimal numbers for a moment flits over the discussion. The definition of a limit allows them to rest in peace. Let ϵ be a positive real number and return again to the sea and its sounds. To say that the distance between the ship and the lighthouse may indefinitely be decreased is only to say that *whatever* the value of ϵ, there will eventually be *some* motion of the waves carrying the boat to a point whose distance from the lighthouse is *less* than ϵ.

Note the crucial double play, which presents itself almost as an incantation or sea chantey: *whatever* the requisite distance, *some* suitable motion can be found. This idea carries over from ships to sequences. S_n converges toward a limit L if *whatever* the value of ϵ, *some* point in the sequence can be found such that *there* and *for points beyond,* the intervening distance is less than ϵ.

I am still midway between metaphor and mathematics, but the full *mathematical* definition requires only a filling in of detail:

S_n converges to L—and there follows a gabble of quantifiers—if *for any* positive real number ϵ, *there is some* value of n, such that *for all* terms in the sequence beyond n, the distance between S_n and L is less than ϵ.

The relevant point in the sequence will in general depend on the desired degree of closeness. In the example of $S_n = 1/n$, the closer the fit, the further along in the sequence the mathematician must proceed. If ϵ is 1/18, the requisite n occurs at S_{19}, which is 1/19. For 1/19, and all points in the sequence beyond, the difference between 1/19 and 0 is less than 1/18.

The definition itself explains its own difficulties. It requires three quantifiers and trades on two inequalities. Experience indicates that these are difficult devices to retain in memory.

Could not the effect of the definition be achieved by trading on a little mathematical body English and a good many solid examples?

Perhaps.

chapter 15

Prague
Interlude

SEVEN HUNDRED MILES TO THE EAST OF PARIS, THERE IS PRAGUE AT DUSK,
a full moon over the Karluv Most—the Charles Bridge—the river
glossy below, swans paddling slowly, Hradcany Castle and the gentle
stone buildings of the Mala Strana in the background. It is there, as a
mist rises from the river, that Bernhardt Bolzano, dressed in the coarse
woolen cloth of a Franciscan friar and dead officially since 1848,
walks late at night, and where he can be heard muttering to himself,
and I mention this queer implausible fact because for the whole of the
time I was in Prague, there *he* was, a gentle, tubby, spectral form.

I had come to Prague to deliver a lecture at Prague University,
where in 1796 Bolzano had entered the faculty to attend courses in theology and philosophy; I found myself lodged in an ancient building

attached to the main part of the university by a common wall. My quarters were palatial, a living room, separate bedroom, bathroom, and foyer. The ceilings were at least fifteen feet high and timbered, the huge wooden beams decorated in what looked like pastels, an effect made unnerving because of the contrast between the solidity of the timber and the fragility of the colors. Later, someone told me that these quarters had been reserved for the daytime use of high party officials. I could imagine a pretty woman sitting in this apartment, braiding her hair, waiting.

Bolzano's spirit flowed freely through the walls, where on occasion it took up residence in the concierge, a mild, inoffensive, middle-aged man; he spent most of his time watching television.

"You lif in California, *tak*?" he asked me one day.

"*Tak.*"

He opened his ledger and began to rifle through the pages.

"Ve had here in April zomvon else from California. Here is. Professor Jacobson. You know?"

It was no one that I knew.

"*Tak*, here he go out in morning, come back for lunch."

An Aspect of Things

Continuity is an aspect of things as rooted in reality as the fact that material objects occupy space; it is the contrast between the continuous and the discrete that is the great generating engine by which the real numbers are constructed and the calculus created. The concept of continuity is, like so many profound concepts, both simple and elusive, elementary and divinely enigmatic. A process is continuous if it has no gaps, no place where the process itself falls into abeyance. The flight of an eagle is an example. The great bird gathers its shoulders, pushes off from a rotted tree stump, lifts into the wind, its wings beating, soars upward on a thermal current, and then, its neck curved downward, folds its wings together and dives toward the stream below. Although in the course of flight the bird does different things, there is no moment when what it does simply lapses so that it *jumps* from one part of its aerial repertoire to another.

We live in disorderly times. Thing seem often *dis*continuous and almost always chaotic. Quantum theory suggests, especially to those who have not studied it, that on some level of analysis, quanta bounce around for no good reason whatsoever. It is worth remembering, if only for the sense of calm that it provides, that the calculus was created by men who looked out on a different world, one in which the great panorama of natural processes were all of them clearly continuous.

In its representation of the real world, the calculus subordinates processes to functions—*that* is its most compelling impulse—and so a definition of continuity must have as its aim a statement to the effect that a function is continuous just in case—and there follows, of course, a moment of confusion in which the familiar fog rearranges itself. Just in case *what*? A first essay at a definition proceeds by imitation. A real-valued function f is continuous if in its behavior it has no gaps. It is here that the imagination endeavors to evoke within a purely mathematical mirror that essential seamlessness so plainly a part of the physical world. And here inevitably the mirror becomes cloudy, returning for the world's bright images something turbid and unclean.

I mention this to my hosts, Professor Swoboda, who is a mathematician, and Professor Schweik, who is a philosopher. We have been walking toward the Vltava River. Both men appear to me to be in late middle age, perhaps sixty or so. They have very similar round, almost globular heads, thinning hair, very sallow skin, and shockingly bad teeth. They are dressed in shabby suits, and their bodies have a defensive weariness.

Swoboda is extraordinarily intelligent. He speaks English in an odd way, appearing to fetch each word he utters from a great distance. Nonetheless, his grasp of English grammar is perfect, and very often he expresses himself not only with precision but with an eerie economy of effect.

We reach the Karluv Most. I am struck by the extraordinary light, in which blue, blue gray, and gray are offset by a kind of smokiness. I wonder if it might be pollution, but there are few cars in Prague and almost none in the central quarter. "The smokiness," says Swoboda, "is the atmospheric effect of the decomposition of the sandstone that has gone into the construction of the bridge itself, an interesting example of an artistic entity making possible the unique conditions under which it may best be appreciated."

"*Tak,*" says Schweik.

"It is the same," Swoboda says, "with mathematics."

A Lecture

That afternoon I walk through the oldest quarter in Prague. The streets are narrow and the houses timbered. It is here that Bolzano was born in 1781. There is everywhere a mood of moral earnestness, as palpable as the lowering color of the sky. Reading a biography of Bolzano published in German, I am not surprised that he proposed to order his life by a principle of *benevolence*. By and by, I wander back to the university to meet with the director of the institute, Ivan Havel. He is a small, energetic, merry man, with gray hair worn in thick curls, gray eyes, and a trim compact body. He is dressed in a well-cut English suit and wears a shirt with French cuffs. He speaks English, which he has acquired at Berkeley, with a considerable Czech accent and lisps as he talks, spraying saliva in every direction; at lunch he is a menace.

Presently Sir Arnold Bergen enters the room. He is an immensely distinguished British pharmacologist, in Prague to deliver a lecture to the Academy of Science. He is perhaps in his mid-sixties, lean, vulpine, his hair covering the top of his head in strands drawn up from one ear; he has a large powerful nose, the thing like a flügelhorn.

We go to lunch at a club said to be frequented by Czech reporters. The food is mystifying. I order dumplings and Bergen, carp. Steins of foaming pilsner all around.

Later, I give my talk in a room with a strange glass blackboard. When I am introduced, Havel says that I am a writer as well as a scientist. This elicits a murmur of approval. There are twenty or so people in the audience. The room quickly grows stuffy, but everyone listens intently.

I speak slowly and distinctly, in the way one does to an audience that treats English as a second language; I am supposed to talk about Tychonoff's theorem, but to my surprise I find myself explaining the elementary calculus to a roomful of mathematicians, re-creating in my own mind the steps that Bolzano took in order to define continuity. For some reason I feel it absolutely crucial to explain how the concept of a limit is applied to functions. No one seems to mind or even notice.

"A function indicates a relationship in progress, arguments going to values. Given any real number, the function $f(x) = x^2$ returns its square, *tak*?"

The solemn serious men nod their heavy heads.

"The image of a machine, something like a device making sausages, is irresistible," I say decisively.

I walk over to the blackboard and with quick strokes of chalk draw a sausage-making machine, or what I imagine looks like one.

"In go the arguments 1, 2, 3, *out* come the values, 1, 4, 9."

There is a snicker from one of the men sitting on the rough wooden pew in front of the class. "Vulgar?" I say. *"Tak,* vulgar." I wipe the palms of my hands on my suit, a gesture that I realize I have never made before. And then I resume.

"As the arguments of f get larger and larger, its values get larger and larger in turn."

Sir Arnold Bergen and Ivan Havel seem fascinated, and I receive the impression that this is material that they have never heard before. Swoboda and Schweik are looking at me intently.

"Now imagine," I say, "arguments coming closer and closer to the number 3, *tak*?"

I walk back to the blackboard and show the men in my audience what I mean, writing, 2, 2.1, 2.2, 2.3, 2.4, 2.5, 2.6, . . . , before the function.

"What then happens to the function? How does it behave?" I ask, realizing with a sense of wonder somewhat at odds with the hard-boiled pose I usually affect, that a function is among the things in the world that *behaves*—it has a life of its own and so in its own way participates in the drama of things that are animate.

"I mean," I say, "what happens to the values of f as its arguments approach 3?"

I look out toward my audience. Swoboda and Schweik are looking at me intently, their faces serene, without irony. It is plain to me that they do not know the answer yet.

"They approach, those values, the number 9, so that the function is now seen as running up against a *limit,* a boundary beyond which it does not go."

Swoboda leans back and sighs audibly, as if for the first time he had grasped a difficult principle. The room, with its wooden pews and narrow blackboard, is getting close.

I say, "The concept of a limit, as it is applied to functions, is forged in the fire of these remarks."

I step back to the blackboard and write:

As x approaches 3, f(x) approaches 9.

Or again:

As x gets closer and closer to 3, f(x) gets closer and closer to 9.

Or yet again:

As x gets closer and closer to 3, f(x) approaches 9 as a limit.

"You see," I say, "the function $f(x)$ *has* a limit at L if, as x approaches some number C, $f(x)$ gets closer and closer to L."

Sir Arnold Bergen mouths the words *closer and closer.*

"The analysis," I say professorially, "proceeds as it has proceeded in the case of sequences; $f(x)$ is getting closer and closer to the limit L if the differences between $f(x)$ and L are getting smaller and smaller, if they may be made small without end—arbitrarily small."

Sir Arnold allows the accumulated tension in his body to collapse. I am tremendously pleased that I have made my point, even though it is a point with which every mathematician in the modern world is familiar.

And then *I* say something that astonishes *me:* "It is when functions are seen in *this* context that the poignancy of the process becomes for the first time palpable."

Several of the men cross themselves.

"In the example of $f(x) = x^2$, the function achieves a moment of blessed release at the number 3 itself; *there* $f(3)$ *is* 9, the process of getting closer and closer over and done with."

"Yes," says Sir Arnold.

Then I write the symbols $f(x) = (x^2 - 1)/(x - 1)$ on the blackboard and rap the board with my knuckles.

"Here," I say, "is another story."

Ivan Havel has let his attention wander, but Swoboda and Schweik are looking at me raptly.

"As x approaches ever more closely the number 1," I say, "$f(x)$, as a few examples will reveal, gets closer and closer to the number 2. It *approaches* 2 as a limit."[1]

I say this dramatically, wiping my hands again on my suit in the same gesture that I find so strange.

"But *at* 1 itself"—I pause dramatically and allow a heavy, meaningful silence to invade the room—"$f(x)$ lapses into nothingness, *tak?*"

Sir Arnold Bergen is frowning again in concentration.

"The function $f(x)$ lapses into nothingness *because*"—I say this word very deliberately—"$(1 - 1)/(1 - 1)$ is simply 0/0. At its limit, this function is *undefined*. The behavior of a function *at* an inaccessible point is expressed or explained by its behavior in a *neighborhood* of that point."

There is another puzzled look from Sir Arnold. I see suddenly what I really wish to say: "The function gets closer and closer to its limit, but, you see, *it* never *reaches* that limit."

Tak, says someone in my audience, *like man to God.*

I say goodbye to Swoboda and Schweik at the metro station. I watch for a moment as they trudge down the street. Their tread is heavy and tired; I notice that they barely lift their feet from the ground.

The Continuous Functions

The light from the sun fills the whole of the sky, and although light may be contained, perhaps, within a room, light itself is everywhere, filling space with radiance. Light is a global physical property as area

---

[1] Given a real number x, this f acts first to square it. Then 1 is taken from the square, the result left to perch on top of the fraction. Down below, f withdraws 1 from x, the work of the function finished when the perchee (on the top) is set off against percher (on the bottom). If $x = 3$, $x^2 - 1$ is 8 and $x - 1$ is 2. And $f(3)$ is 8/2 or 4. At $x = 1$, both $x^2 - 1$ and $x - 1$ are 0. The fraction 0/0 that results has no mathematical meaning and exists entirely as a symbol of nothingness. At $x = 1$, f is undefined. The number 1 marks the spot where a black hole exists, a great emptiness. Everywhere around the number 1, like peasants tilling fields on the slopes of a volcano, the life of the function continues. As x approaches 1, $f(x)$ approaches 2. And yet at 1 itself, nothingness predominates.

is a global mathematical property. Continuity, by way of contrast, is a *local* concept; it is defined at a point, and it makes a claim about the behavior of a function in the here and now, *this* place and *this* time. In order to coax a concept from the alembic of words, the mathematician's eye must linger at a point x and a *neighborhood* of points around that point.

I am struck again by the depth of contrast between a place or point and a neighborhood when at the end of another day I return to the Karluv Most, the center of Prague, the fantastically beautiful stone bridge across the river. At the far end of the bridge, an American couple are singing songs from the sixties, "Michael, Row Your Boat Ashore" especially popular. They are singing the song enthusiastically, strumming their guitars with great thrashing strokes. I wonder what impels these young people to leave Burbank or Pasadena in order to sing along the Karluv Most. I know the answer myself. Almost every European city offers the illusion of finely segmenting the manifold of experience, so that to an American sensibility Prague appears as dense with possibilities as it is physically reticulated by small streets, urban alleys, footpaths, little walkways.

A neighborhood is the place where a point finds itself. The size of the neighborhood is not crucial to the concept of continuity; the assumption is that it is small. If $x = 3$, a neighborhood of 3 might be those numbers within the interval between 2.5 and 3.5. The neighborhood is signified by brackets, as in [2.5, 3.5], meaning those numbers between 2.5 and 3.5—*all* of them, including the numbers 2.5 and 3.5 themselves. The relevant interval is *closed*. It contains its endpoints as a designated city street might contain the houses on either end of the street, as the designated city street in which I am lodged, the concierge reminds me, *does* contain the historical houses on either end, protecting them in a closed neighborhood of houses around the university.

The idea of a closed interval is so obviously half of an idea that like a fraternal twin it suggests at once its other half. On Celestina Street, the historic houses at either end are included within the street's historic ambit; elsewhere in the city things are different. "*Tak*," the concierge remarks, "up to end of street, piple go, but not last houses." Such intervals are *open,* the good stuff lying *between* the margins.

Round parenthetical brackets signify open mathematical intervals, as in (2.5, 3.5), the numbers between 2.5 and 3.5 swimming endlessly up to those now inaccessible margins at 2.5 and 3.5.

Defined in a neighborhood of *a*, a function *f*—

What?

—it means that the function *f* makes sense at every point in the neighborhood; it means that points in the neighborhood are included in the function's domain, its theater of operations. The function $f(x) = x^{1/2}$ is *not* defined in *any* neighborhood of 0. Negative numbers lack square roots, and the neighborhoods of 0, whatever they are, contain negative numbers.

Where was I?

Yes, yes. Defined in a neighborhood of *a*—I am continuing with the definition—a function *f* is continuous at *a* if the limit of *f(x)*, as *x* approaches *a*, is *f(a)*. This dense declaration goes over elegantly to symbols, a first indication, if any were needed, of the matchless power of mathematical notation. The function *f* is continuous at *a* if

$$\lim_{x \to a} f(x) = f(a).$$

It would be hard to improve on the simplicity of this formulation, its uncanny concision, and hard, too, to improve on its intuitive appeal. A function is continuous *at a* if in its approach *toward a* its behavior is uninterrupted, if in its approach toward *a* the damn thing simply slides.

Still, this is to describe things from the outside, in terms of metaphors and images. From the inside, the definition of continuity unwinds itself into a three-part declaration: If *f* is continuous at *a*, *f(a)* itself must make sense, with *a* in the domain of *f*. This is a condition of *intelligibility*. The limit of *f(x)* as *x* approaches *a* must in turn be *out there*, alive and quivering. A condition of *existence*. And that limit must in turn be a value of *f*, the function not only approaching some limit or other but merging rapturously with the limit at *a*. *Consummation*. The concept of continuity thus resolves itself into the triple concepts of *intelligibility, existence,* and *consummation.*

Continuity Gone Wrong

Inasmuch as the definition of continuity is given in three clauses, discontinuities arise in three ways. The simplest of all discontinuities occur owing to a lapse in definition. At $x = 4$, the function $f(x) = (3x - 12)/(x - 4)$ goes bad; it is at 4 that $f(x)$ suffers a discontinuity. And ditto for countless other functions in which division by zero figures as a permanent threat, a black possibility.

When my meal in a Prague restaurant comes, it consists of two dingy gray slices of cutlet over a mound of French fries. It is designed to be *medallions of bifstek,* my waiter explains, over *champignons.*

Yes, well, where are those champignons?

My waiter looks down at the plate and pokes at the meat with his finger. *"Nikt da,"* he says mournfully after finally turning my cutlet over. He had discontinuities of the first kind in mind.

A second kind of discontinuity arises when the required limits fail to appear. Defined as it usually is, the function $f(x) = 1/x^2$ gets larger and larger as x approaches 0. $f(1/2)$ is 4, $f(1/4)$ is 16, and $f(1/6)$ is 36. At 0, f appears to undergo an ordinary discontinuity: $1/0$ is nothing. Yet nothing prevents the mathematician from simply *saying*—it is a perk of power—that at $x = 0$ the value of this function should be 1 instead of $1/x^2$. The function that results is yet discontinuous at $a = 0$, but *not* because $f(a)$ is undefined. On the contrary: $f(a) = 1$, and this by definition. The discontinuity occurs because the limit of $f(x)$ as x approaches 0 does not exist, the function simply getting larger and larger as x gets closer and closer to 0.

Yet another kind of discontinuity arises when the limit of a function f at a is not $f(a)$ at all, but some other number entirely, an alien interloper. The function $f(x)$ has the value x^2 everywhere *except,* I now assert, at 0. Its value there happens to be 1. This is a stipulation. I am creating this function out of thin air. Now as x gets closer and closer to 0, $f(x)$ plainly approaches 0 as well: $(1/4)^2$ is $1/16$, but $(1/6)^2$ is $1/36$, a much smaller number. Yet at 0 itself, $f(x)$ indignantly bounces back up to 1 in a startling display of personal perversion. The limit of this function at $a = 0$ simply does not coincide with $f(a)$.

At the Kafka exhibit in the heart of Prague, there are pictures of

Kafka on the wall and generous excerpts from his novels. I look at all the exhibits. There are many interesting pamphlets for sale and a number of books about Kafka and Prague. I approach the young woman at the counter and ask for something *by* Kafka. She shrugs her plump shoulders. "Iss sorry," she says. *"Für diesen Bücher müssen Sie nach Deutschland."* For that you must go to Germany. All that activity points in one direction; the limit lies elsewhere.

As I am rehearsing these ideas on the Karluv Most, I find myself casting about for an illustration, something that shows the meaning of continuity. *Tak,* Bolzano's spirit tells me. *Look.*

At the other end of the bridge, two Czechs are playing old Slavonic folk tunes, one on the oboe, the other on the guitar. They are fine musicians both, the oboe and the guitar making for an unusually successful combination. The melodies probably go back in racial memory to the Middle Ages. No one has to ask what the songs are about. As I am listening, I turn and look upstream at the Vltava. A thin, elegantly long cruise ship has just emerged from underneath the bridge and is gliding upstream, its lights subdued and carefully modulated, the very opposite of the garish Parisian *bateaux-mouches* on the Seine; as it glides noiselessly along, a sinuous trolley consisting of three linked cars begins to cross the heavier industrial bridge upstream from the Karluv Most, the sound it makes entirely muffled by distance, so those two enchanted objects, the ship and the tram, appear to be moving silently toward an imaginary point of intersection where their separate worlds of light will merge fantastically into a luminous and limpid starburst.

Lives of the Concepts

No matter the concepts, the lives that they lead are connected with the lives that *we* lead, continuity, especially, being withdrawn by the mathematician from the real world itself.

Later that day, after I had shaved and showered, I walked to the Stare Mesto, the old city, where I watched the famous Baroque clock turn the hour. Then I wandered over to the ancient Jewish quarter: tours of the cemetery were being sold for a few kroner. I stopped and observed a pair of young men press brass coins on an old-fashioned coin-stamping machine. After lunch, walking toward the river, I saw a sign: Dr. ———; there followed a complicated Czech name; the man was a specialist in Calanetics, an American weight-loss system having something to do, I remembered, with cellulite. I thought of an exercise room in which a number of Czech women dutifully were lifting their meaty thighs. "Vun, two, and up, ladies," says the Czech Calaneticist.

I crossed the river and by and by, I came upon a pastry shop. There were only three trays on the shelves, and the pastry was waxy and abhorrent. Nonetheless, there was a long line. One woman waited by the cash register staring into space and making change with enormous deliberation; another stood behind the counter and took orders, writing things down laboriously on a square of oil paper. Suddenly she dropped what she was doing and scuttled to the back of the shop. After a few minutes she emerged with a new pastry tray containing things that looked like purple crumpets. She began leisurely to rearrange the items on the shelves, first taking out one tray of pastries, and then the other. Business came to a halt. I stomped out of the store, pastryless, fuming with inexpressible indignation.

Walking once again toward the river, but from the other side, I was struck by what were obviously very old large globes covering the street lamps. They had been segmented by black wire or paint in order to form a series of interlocking pentagrams. The effect was of unusual artistic interest, the globes suddenly appearing faintly fleshlike and swollen, the contrast to the geometrical severity of the design striking and elegant.

I mention these details of life as it was lived one afternoon in order to suggest that the simple act of moving continuously from one place to another on the surface of the earth has a rich but hidden structure, one that requires an act of meditative *mathematical* attention before its essence is revealed. The central European afternoon and evening were filled with things seen, streets crossed, sights admired, but there is at least one simple description of what I did that compels the day's accidents to defer to their essence. *I walked from one side of the Vltava River to the other, crossing the river by the bridge*. The map depicts the path that I took:

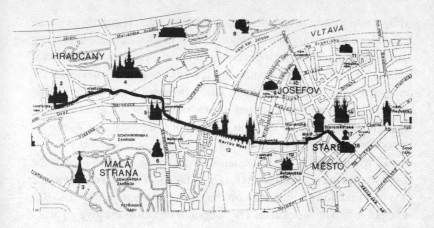

Described thus, my trip has already receded along a corridor of abstraction, with what I *did* replaced by a stylized *record* of what I did.

That enchanting map now disappears in favor of a Cartesian coordinate system, my path between places supplanted by a continuous curve:

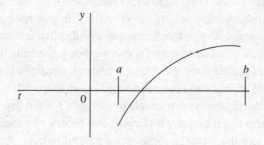

Cartesian coordinate system
with continuous curve

These changes are more than cosmetic; they mark an enlargement in abstraction. The map's adherence to the real world is giving way.

The distance between what I did and how the mathematician describes what I did is now about to widen once again. A continuous function is continuous at a point; continuity is, like position, a local concept. But a function may well be continuous at a point and continuous, as well, at *every* point in a closed interval [*a*, *b*]. Such is continuity *on* an interval, a concept cognate to continuity itself. This definition does not cancel the connection between continuity and a point; it does not violate the local character of continuity. It means only that what holds for one point holds for others, the function behaving very much like an amiable longshoreman stopping at one waterfront bar after another, where at each, he hoists a round. His actions are local, but as the afternoon wears on, he covers the neighborhood.

In what follows, the Cartesian coordinate system abides, but the curve wandering across the *t*-axis is now taken—*I* am taking it—as the face of a function *f* continuous on the closed interval [*a*, *b*]. At *a*, *f*(*a*) is negative and at *b*, positive. These double suppositions give an articulate, a precise mathematical voice to the scenario already sketched, but behind the mathematics there is always the link backward to living life. Continuity on the interval simply affirms what I already know: that in walking around aimlessly I also walked around continuously, walking a member of the great interracial tribe of continuous processes. The requirement that *f*(*a*) be negative and *f*(*b*) positive simply duplicates on the coordinate axis my ramble *across* the river, the river anticipating the coordinate axis to come, and the coordinate axis standing in for the river. And, indeed, there behind the symbols *I* may be spotted, a somewhat reclusive middle-aged man crossing a bridge in a faraway country. Yet the mathematical description accomplishes something that my impressionistic account does not. This Cartesian coordinate system could be any spatial expanse and the continuous curve any continuous curve, so that the coordinate system stands in not only for the trip that *I* took but for any trip *like* mine, any linking of two places by means of a continuous activity that crosses somehow a natural barrier. And here the calculus reveals, perhaps for the first time, the extraordinary reach of its concepts, the wide-ranging generality of its attitude.

The Sign of Three

"The Shepherd in Virgil," Dr. Johnson remarked in his famous letter of rebuke to Lord Chesterfield, "sought for love and found him a native of the rocks." The obscure reference is to Virgil's *Eclogues*.

Dr. Johnson wished to call vividly to a classically trained mind the fact that love may be the source of misery and desperate unhappiness. I have always been haunted by this line; like so many of Johnson's aphorisms it points to a profound puzzle presented by our own emotional states—that despite their familiarity, they are infinitely mysterious. One paradox leads to another. The concepts of mathematics, despite their *un*familiarity, are infinitely accessible. At their deaths, those who have minded mathematics will have known the continuous functions better than the crooked human heart. That so abstract a consideration should in the end be so lucid is a source of wonder.

A great mathematical property reveals its identity in terms of the theorems that it makes possible. In the case of continuity, there are three subtle and powerful theorems that with a hot white light illuminate the inner nature of the idea.[2] They are all of them theorems about f itself, and they are all *global* in the sense that they reveal aspects of the continuous functions that hold for the whole of the interval on which they are continuous; such theorems are typically very powerful and very hard to prove. These theorems are about f, but they are also about the processes that f and functions like f represent; and so they make a claim about the composition of the world, its true, correct, and inner nature. With the theorems established and in place, the continuous functions are forever marked by the properties of *boundedness, betweenness,* and *maximality*—the Sign of Three.

The property of *boundedness* is lifted directly from life, where it plays a modest but nonetheless unmistakable role. In walking through the Stare Mesto and then across the river, there is some street on the far side of the river that marks a point *beyond* which I did not go. And this the map itself reveals: *any* of the streets beyond the street where I turned back toward the river might serve as a boundary. As it happens Loretanska Street is a natural boundary to my ramble. Typically, it is a street of this sort, the tourist or the traveler discovers when it is too late, that marks the threshold of unlimited enchantments.

---

[2] The proofs of these theorems are generally thought to fall outside the domain of the calculus. They are in any case very subtle. A proof of an intermediate value theorem is given in appendix 1. It is very subtle.

"*Tak,* you go past Loretanska?" the concierge inquires when I return.

"No," actually I didn't.

"*Tak,* iss too very bad."

Taken in the context of my walk, boundedness appears as an obvious but unremarkable feature of experience; but boundedness is, in fact, the first of the great properties of continuity, a property possessed by *all* the functions continuous on an interval. Let the requirement that a function be continuous on an interval lapse, and boundedness may lapse as well. A definition now follows. A function is *bounded* on an interval [a, b] if some number exists that trumps the function—in symbols, if there is a number N, such that $f(x) \le N$ for every x in [a, b]. There is nothing remarkable about N: it need only be bigger or at least equal to the best the function can offer.

It is a theorem, and so a fact, a part of the world's immovable furniture, that *if f* is continuous on the interval [a, b], then it is bounded there as well.[3]

If boundedness appears unremarkable, *betweenness* is more unremarkable yet, its role in real life often overlooked if only because of its overwhelming obviousness. In going from one side of

---

[3] The function $f(x) = 1/x$ is continuous everywhere but at 0 on the closed interval [0,1]; at 0, it is undefined. This function is unbounded on [0,1]. Note that as x approaches 0, $f(x)$ gets larger and larger, but that as it approaches infinity, $f(x)$ gets smaller and smaller. The picture shows the behavior of the function:

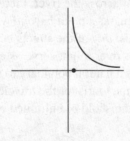

Face of $f(x) = \frac{1}{x}$

the river to the other, I crossed—it was *necessary* that I cross—the sluggishly moving river itself. It is in general true that to get from here to there, I must pass through the points *between* here and there. Even those sent from the USS *Enterprise* to the planet Zork by transporter on *Star Trek* pass as disembodied rays (or whatever) through the points intermediate between the USS *Enterprise* and Zork. So strong is our natural assumption that processes in nature are continuous that we would be intellectually dismayed by an episode of *Star Trek* in which the captain simply appeared in a distant place without passing through any intermediate points. The continuous processes are marked by their commitment to *intermediate* values, the very essence of continuity having something to do with betweenness. But how much better the point is put in mathematics than in the rambler's loose-knit prose. A number x is *between* two other numbers y and z if it is greater than the first, less than the second. The point again in symbols: $y < x < z$, those symbols systematically revealing themselves as the infinitely elegant, infinitely useful characters that they are.

Strolling across the Karluv Most, it is the river below that marks the point between banks; one of the charms of rivers and bridges, I suppose, and the source of their irresistible attraction, is simply the fact that within very little space they offer the impression of changing dramatically the circumstances of one's body. Back on one side of the Vltava River, I am over here; on the other side, over there. This will seem a childish pleasure only to those immune, as I am not, to bridges and their charms.

To capture the shimmering river in silvery symbols, the mathematician assumes that $f(a)$ is negative and $f(b)$ positive, so that 0 is between $f(a)$ and $f(b)$, the whole of that river compressed now in 0 to a single number. And then it obligingly affirms, the second of three great theorems on continuity, that *if f* is continuous on $[a, b]$ and *if* 0 lies between $f(a)$ and $f(b)$, then there is—there *must* be—some number s in $[a, b]$ such that at s, f is itself 0. The number s marks the spot where the river flows, the spot as well where the humped and ancient bridge arches over the water; the spot in Prague intermediate between two banks of the river, two far-flung neighborhoods, two worlds; and the spot, finally, where f, the stand-in for all this loveliness of life, itself takes on an intermediate value at 0.[4]

Maximality is the last of continuity's great properties, the third of the Sign of Three, and the property least connected to experience. The continuous functions are bonded to one another by the fact that on closed intervals they take—they *must* take—their maximum and minimum values. Yet a connection of sorts exists between experience and mathematics. My ramble is bounded, I remarked, if there is some street beyond which I did not go. It is a *maximal* ramble if at some point in my walk *I* have gone as far as *I* can go. Note the subtle difference, and note, too, how unaccountably difficult it is to express in ordinary English. Boundedness refers to a street beyond which I did not go: it is a barrier *I* need not have reached. It suffices that the boundary is simply *there*. Maximality refers to a point in my walk that *I* myself reached, the outermost point that *I* got to. The stress is now on the places that *I* reached and not on the barriers that I did not cross and may not have even seen. Loretanska Street served as a boundary for my walk; I never reached the street, never saw the castle beyond. *I* got as far as Vlasska, a dark medieval street, as I recall, in which someone in an upstairs apartment was solemnly playing a violin.

---

[4] Suppose that $f(x)$ is -1 whenever x is greater than or equal to 0 but less than 2, i.e., $0 \leq x < 2$; but that it is 1 whenever x is greater than or equal to 2 but less than or equal to 4, i.e., $2 \leq x \leq 4$. This strange function jumps over the x-axis at 2 itself, violating the conclusion of the intermediate value theorem. But, of course, f is not continuous at 2 either. The picture shows what is at issue:

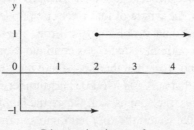

Discontinuity at 2

"*Tak,*" remarked the concierge, "you go to Vlasska? Once I knew girl there."

I looked up, interested.

The concierge pantomimed someone shooting a rifle. "Germans, they leave."

Loretanska is the boundary; but Vlasska—that is the maximal point to my walk. And to someone else's life, it would appear, the endeavor to explain a fact of mathematics revealing a connection to sadness and to sorrow.

Boundaries and boundedness refer to numbers, maximality to the behavior of a function. A function has a *maximum* value on an interval if there is some point where—at *that* point—the function is greater, or at least equal to, its incarnations or identities at all other points in the interval. Symbols may by now be seen a sweet relief. The function f has a maximum at y if $f(y) \geq f(x)$ for every vagrant x. And the last of the three theorems affirms that *if f* is continuous on $[a, b]$ there *is* a number y such that $f(y)$ *is* greater than or equal to $f(x)$ for every x in $[a, b]$.[5]

It is in this fashion, then, that continuity reveals itself as a concept. The functions that are continuous on a closed interval are bounded there, they take intermediate values, and they achieve their maximum

---

[5] Suppose the function $f(x)$ has the value x^2 if x is less than 1, but if x is greater than or equal to 1, $f(x)$ drops to 0, where it forever after remains. This abject function is bounded on the closed interval $[0,1]$ (by 1 itself), but it does not take its maximal value on this interval. The natural thought that $f(1)$ must be greater than any other value of f comes to grief on the realization that $f(1)$ is 0 and so *less* than the other values clambering upward. Again, a discontinuity at a single point (1, in fact) destroys the conclusion of the theorem. And again a picture shows the moral of this message better than the message shows the moral:

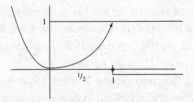

Taking no maximum

(and minimum) values. And this may be seen at once on the map; it may be seen as well in the more austere context of a Cartesian coordinate system:

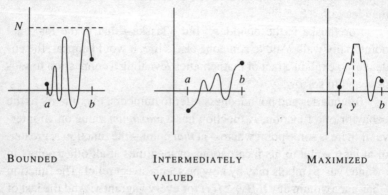

BOUNDED   INTERMEDIATELY   MAXIMIZED
                           VALUED

The Sign of Three

It is to Bernhardt Bolzano that the modern concept of continuity is due; he was not like Cauchy a great powerhouse of mathematical thought; his intelligence was delicate but prophetic. Standing before the ornate astronomical clock in the town square, or passing reverently before St. Vitus from Novy Svet, or simply walking over the Karluv Most, he seems to have had a gentle gift for divining the future of thought, a gift that time bestowed but compromised by depriving Bolzano of the degree of mathematical mastery needed successfully to develop his ideas. He was interested in the foundations of the calculus, seeing that the calculus must purge from its concepts the last of the infinitesimals; and he was offended by shabby standards of proof. Yet Bolzano often blundered; he missed the counterexamples that other mathematicians spotted; he left his ambitious program for the philosophy of science incomplete. He wrote volumes, but they remained in manuscript. And yet the evidence is incontestable: it is there in his writings on every page: *He saw, he saw*.

I encountered Bolzano on the Karluv Most on my last night in Prague. I had been aware of him all evening, a round figure in a brown monk's cowl, walking ahead of me, muttering. As we reached the middle of the bridge, he turned and drew back his monk's cowl. I looked at his open, honest face, with its thick stubble along his chin and cheeks. For a moment I stood like that, embarrassed as one is always embarrassed in meeting the dead. Then he said what I knew— what I had always known—he would say. It is what the dead always say, and it is the only thing they say.

"Remember me."

The
Intermediate
Value Theorem

The intermediate value theorem expresses a fundamental property of continuity. It would be dismaying if this property simply *appeared* among the continuous functions, of no more ultimate significance than red hair among the Irish, but there is a deep connection between the properties of the continuum and the chief property of continuity, an abstract dependence that reveals a throbbing and thus a living connection between ideas.

The intermediate value theorem affirms that so long as f is continuous on an interval $[a, b]$, there is going to be a number somewhere—

And this is the first great step undertaken by the proof: its initial commitment is to *find* a number. Not any number, of course. It must be within $[a, b]$, and it must have the straightforward property of sending f to 0.

The exercise of finding numbers calls the Dedekind cut to mind, that and Dedekind's axiom. Dedekind's axiom brings a number into existence when a cut is made amongst the real numbers. Very well. Suppose that the numbers are divided into two camps A and B.

A is a collection of numbers, a great swarm of them. *Assume* that among them there is a number y with the following properties. While less than or equal to b, y is greater than or equal to the other numbers in A. These twined conditions are joined in one symbolic statement: $x \leq y \leq b$. And what is more, $f(y)$ is less than 0. In symbols again, $f(y) < 0$.

Like A, B is a crowd of real numbers, but a crowd with a difference. It contains *no* number y satisfying the twined conditions that $x \leq y \leq b$ and $f(y) < 0$.

A and B, disorderly as they are, define a cut among the real numbers; and a cut having been convened, it follows from Dedekind's axiom that there exists a number s such that the numbers less than s belong to A, the numbers greater than s to B.

In particular, the number $s - 1/n$ belongs to A because it is less than s, and the number $s + 1/n$ belongs to B because it is greater than s, and this is true whatever the natural number n.

Now as n gets larger and larger, $f(s + 1/n)$ approaches a limit, and because f is continuous

$$\lim_{n \to \infty} f\left(s + \frac{1}{n}\right) = f(s).$$

Recall that B contains no number y such that $f(y) < 0$ if $x \le y \le b$. This means that for any choice of n, if $f(s + 1/n)$ is less than 0, $s + 1/n$ must be less than other numbers in B. But as n increases, $s + 1/n$ decreases. With the Devil to one side, there is only the deep blue sea to the other; it follows that

$$f\left(s + \frac{1}{n}\right) \ge 0,$$

from which it follows that $f(s) \ge 0$.

One half of the proof is now complete. The other half reverses the intellectual steps undertaken, showing that in virtue of the construction of A and the definition of continuity, $f(s) \le 0$. Yet a number simultaneously greater than or equal to 0, and less than or equal to 0, must, in fact, *be* 0. This is the desired conclusion.

This is a very lovely argument, but a very difficult one as well, the difficulty as much a matter of logic as of mathematics. The reader who takes it all in stride has missed his or her calling.

Beyond the details, there is the larger drama of connections achieved between ideas. A proof is a stylized literary exercise, one congeneric now with an epic, at other times with a quatrain, still later with a lyric; in this proof, something is wanted and something is found, its form that of a romance and the pattern an old one of absence and redemption. But the main message conveyed by the proof is one of action at an intellectual distance; the number at which f is 0 is charmed into existence by a spell cast by two widely separated concepts, continuity on the one hand and severability on the other. This provides much needed reassurance that there is a connection in the calculus between concepts.

The Limit
of a Function

The concept of a limit, like a Greek player, wears two slightly different masks, one turned toward sequences, the other toward functions; but only *one* essential idea lurks behind the masks, one superbly utilitarian instrument.

A real-valued function f is given, and as its arguments run up against some fixed boundary, the function converges toward a limit. The definition of a limit must at once give expression to an important idea *and* control the delicate adjustments necessary in order to bring quantifiers and inequalities into alignment.

A limit is, of course, a *number;* and whatever the limit, a function converges toward that number if the distance between the values it delivers and the number itself may be made arbitrarily small. And so the definition must somehow say that *whatever* the positive real number ϵ, at some point in the scheme of things $f(x)$ must be within ϵ of L.

So far, the definition simply tracks the remembered clause in the definition of a limit given in the case of sequences. A sequence converges toward a limit if all points beyond a certain point in the sequence fall within an arbitrary distance of the limit. It is this sense of *all points beyond a certain point* that must be captured anew in the case of functions.

The convergence of functions embodies a double motion: the function getting closer and closer to a limit as its arguments get closer and closer to a fixed point. As x approaches 3, $f(x)$, recall, approaches 9. The definition of a limit must control and properly coordinate these two mental movements.

The device is similar to the one adopted in the case of sequences. If ε is any real number greater than 0, then δ is the same, a real number greater than 0. This number δ will mark out those of the function's arguments that satisfy the requirement that *there* the function's values fall within ε of *L*.

The crucial clause in the difficult definition is the one that establishes the asymmetrical dependency between ε and δ. Matters are handled by means of quantifiers—the mathematician's marker of quantity. For any ε, the definition begins, there is a δ, meaning that whatever the choice of ε, some suitable choice of δ can be made. It is the controlling universal quantifier that determines the specific existential choice of δ. The body of the definition then says that if a function *f* converges toward a limit, as *x* approaches a value *c*, then however small the distance ε that is chosen, there is some δ such that for all arguments that are within δ of *c*, *f(x)* is within ε of *L*.

Or as the pros put it: The function *f(x)* approaches the limit *L* as *x* approaches *c*, if for any ε, there is a δ, such that *if* the *positive* distance between *x* and *c* is less than δ, then the distance between *f(x)* and *L* is less than ε.

Note a crucial point: the definition appeals only to positive distances between *x* and *c*. With what happens at *c* itself the definition does not concern itself; indeed, a function may approach a limit without ever reaching the limit or even being defined at its limit. The function $f(x) = 1/x$ provides an example. As *x* approaches infinity, getting larger without end, *f(x)* approaches 0, but at 0 itself, the function is undefined.

A very helpful picture shows what is at issue:

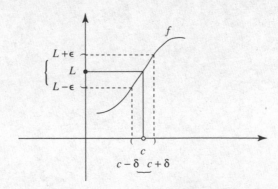

To say that the distance between *f(x)* and *L* is less than ε is to say—the meaning is identical—that *f(x)* lies somewhere between *L*+ε and *L*−ε. This is a point that often maddens students and readers, but it is only a matter of elementary algebra. Say that $L - ε < f(x) < L+ε$, as the picture indicates. Then subtract *L* from each part of the inequality, yielding $-ε < f(x)-L < ε$. It is easily shown from the inequality that $|f(x) - L| < ε$, where $|f(x) - L|$ is the *absolute*

value of the difference between $f(x)$ and L. But another way of conveying the absolute value of the difference between two numbers is to speak of the distance between them, which is what the definition does.

These are very subtle points whose elaboration required the work of centuries—and this by mathematicians of the highest order. The reader inclined to defer the work required fully to appreciate the definition, and to repair back to the text, has my sympathies. He or she will yet be able to appreciate the forward movement of the calculus.

chapter 16

Memory
of Motion

As a child I was fascinated, as all children no doubt are, by a handheld device in which by applying pressure with my thumb to a series of stiff pages engraved with painted rabbits, I could actually manage, once I had acquired the requisite trick of releasing those pages evenly, to prompt that bewitching bunny to lift off from its crouch and, by means of a series of jerky but nonetheless vivid and occasionally spectacular hops, cross a small stream bordering a flower-strewn meadow, whereupon it stopped, whiskers frozen in midtwitch, until by flipping the cards backward, a considerably more complicated maneuver, I reversed its course and made the rabbit cross the stream and return to the place whence he came.

Many years later when I was studying the calculus, the pages of a dog-eared text propped between the pages of the New York City budget, it occurred to me in one of those vivid insights, prompted, I am sure, by anxiety lest I be discovered studying mathematics instead of disbursing welfare, that the secrets of speed lay there in those cards, so that the least and most sophisticated of all human creations, those cards and the calculus, turn out in the end, as one might expect, to be connected by a mystic thread of memory.

Speed is a part of the great wheel of concepts that the calculus treats, but speed has thus far emerged as the ratio of distance to time, this an *average* concept, one contingent upon two different times and two different places. The result is a nonetheless useful formula:

$$\textbf{AVERAGE SPEED} = \frac{P(t_2) - P(t_1)}{t_2 - t_1},$$

or what comes to the same thing:

$$\textbf{AVERAGE SPEED} = \frac{\Delta P}{\Delta t},$$

where ΔP and Δt are simply a shortened way of conveying the difference between two positions and two times, the Egyptian-looking ΔP and Δt read aloud as *delta P* and *delta t*. But however useful the formula, it remains resistant to the scheme undertaken by the calculus. There is no obvious way in which average speeds may be distributed to each of the sparkling instants within an interval, and no way as a result to represent speed as a function of time.

Standing on the pool's high diving board in summer somewhere in New Hampshire, a young girl in a striped yellow bathing suit crouches forward, her antelope knees pressed together and her hands poised like the prow of a ship. Time slows to a crawl and then stops, fixing the diver in the agony of last-moment anxiety; then after a final moment of indecision, my own heart signaling the resumption of time, she topples forward, still spastically frozen in her absurd comma-crouch. Having lost all hope in that first perishable second of achieving a head-first position, she takes up the precious part of another second in more or less rotating her torso slightly in the shedding sunlight, and then with her arms still hopelessly extended, and a goofy grin etched enduringly on her snubbed and childish features, she continues her descent, her entrance into the pool signaled by a resounding flop, the flop smarting so much more from the high board than the low simply because so much more time has elapsed before the bright blue water of the pool has come rushing up to meet that tender tummy.

The brilliant blue sky of a remembered summer's day, sunlight glinting off the water, and the dive and the diver's forward fall, *they* all combine to convey an impression of seamlessness, the dive one of those processes that take place without breaking into discernible parts. The genius of the calculus is to represent such seamlessness by a sequence of discrete stages, one extended to its limit. And strangely enough, the recovery of that sequence from the forward and downward sweep of fluid and remembered motion is easily accomplished. Looking backward along the corridor of memory, I see her standing there, frail brown shoulders hunched and rounded, and quite on its own, the spool of memory slides into its socket and the delicate diver's cheerful belly flop proceeds yet again to unwind itself; but here is the subtle, unnoticed, easily missed fact so effortlessly superimposed on all that descending loveliness—that beyond the ordinary if vivid series of English metaphors that serve to set the spool of memory spinning and reveal the diver's descent as a single continuous arc, there is an equally plausible and no less evocative series of *mathematical* metaphors that describe that arc as an agglomeration of discrete stages. Proceeding downward, her body a touchingly tense and contracted ball, the diver falls first for a certain distance ΔP in a certain time Δt; she then falls a further distance ΔP in a further amount of time Δt, until, meeting the water in that aching flop, she falls no further at all, paddling limply toward the pool's silvery ladder, no worse for wear at twelve, exultant at actually having come down the hard way from the high board.

And the ineffably subtle point is that in talking, by means of the indefinitely elastic ΔP and Δt, of *certain* distances and *certain* times, the mathematician has imposed a discrete schema over what until now has made an appearance as a continuous process. It is the imposition of this schema that dramatically reveals relationships between time and distance that until now have been sealed in seamlessness, hidden from inspection. Objects in free fall on the surface of the earth, Galileo's law affirms, fall under the control of one and the same law; although they may sing in their chains like the sea, the drooping child in New Hampshire and the dropped ruby in Tuscany descend to the surface in a way determined by the fact that distance is a simple function of time. Quite without mathematics, what is there to say of their speed, those drooping, dropping objects? Only that it is the ratio of the distance covered to the time elapsed, which in the case of memory's delicate diver is measured by the height of the high board—twenty-five feet or so—as contrasted with the time that has elapsed between the first forward topple and the last resounding flop. *One* continuous act results in a *single* assessment of speed. But the imposition of discrete distances and times upon the act reveals a wealth of additional descriptive detail. Having fallen a distance represented by ΔP in a certain time Δt, the diver's average speed, or rate of change, is simply the fond familiar ratio $\Delta P/\Delta t$— but the ratio now taken is of *this* distance to *this* time, the diver's forward fall halted in midflop at Δt, where Δt is some time short of her entrance into the water. There are now as many assessments of speed as there are intervals Δt of time. And of these, there are infinitely many.

The world revealed by the senses recedes, a new world of relationships and dim mysterious connections slowly coming into consciousness. As time *expands,* for example, Δt getting larger and larger, distance expands as well, ΔP getting larger and larger, and this commonsensical remark simply reinforces what our senses would otherwise disclose: the longer you fall, the further you go (and the harder you fall, the voice of experience chimes in). But as time expands, *speed expands as well,* the ratio of ΔP to Δt itself becoming progressively larger. Objects falling freely undergo *acceleration.* To be sure, acceleration is something felt by the human body and so a familiar fact of falling, as when the elevator suddenly seems to drop from underneath our feet, but the English language lacks almost entirely the resources to describe acceleration in terms that go beyond the familiar *faster and faster*. The variables of distance and time do better. They concentrate the attention on the essentials at the expense of the details. They allow a precise statement to be enforced so that the *fact* that as time expands speed increases now admits of an unambiguous amplification: as Δt increases, the amplification affirms, so, too, $\frac{\Delta t}{\Delta P}$.

The expansion of time under the scheme just scouted corresponds to time's variable but forward motion. It is not time, of course, that is proceeding in those discrete stages marked by Δt, but rather the mathematician's *attention,* which by an act of rarely used mental muscles is being forced to focus on the flow of time, breaking that otherwise smooth stream into stages. But if the discrete expansion of time is a mental motion, it should be possible by means of an oblique alteration of the mathematician's mental gearing, to reverse course in order to contemplate time's *contraction.* As time *shrinks,* Δt becomes smaller and smaller, the mathematician's attention confined to ever narrower intervals of time. It is with the contraction of time that the first formal concept of the calculus—*the derivative of a real-valued function*—signifies its imminent Technicolor arrival on the great mathematical screen.

The position function is a living record of where an object has been and where it is going and so a means of conveying an object's history, its associations in the world. Galileo's law of falling objects gives that function one familiar form: $P(t) = ct^2$, distance varying as the square of time. It is this function that brings about a tight, quantitative connection between what an object is doing—falling freely—and the time in which it is doing it. The law is universal, holding indifferently for objects whatever their mass, and so in analyzing speed in this context, the mathematician and the novelist part company, the novelist left looking forlornly at that honey-hued, frail-shouldered child, and dreaming of best-sellers to come, and the mathematician reducing those limpid details to the enduring stolidity of a clock and a tower, abstractions signifying time and distance.

The world is emptied now of everything save the universal clock that is ticking everywhere and to which the calculus appeals: an enormous oval, its single spidery black hand progressing around its face, moving from 0 to 1, from 1 to 2, from 2 to 3, and from 3 to 4, the numbers corresponding to numbers on the number line; and a tower one hundred feet in height, clock and tower standing in for clocks everywhere and towers anywhere.

The moment is the infinitely pregnant present. The hand of the universal clock stands at 0. At that very instant, nothing has yet happened, no time has elapsed, no distance achieved, and so there is no speed to speak of. This trite fact may be conveyed by allowing time to roll forward by a certain extent Δt, and then contract, so that average speeds are assessed over shorter and shorter intervals of time. If time could actually be rolled backward, the result would be the familiar revision of those amusing movies in which real life is reversed, the kiss separating from the lovers' lips, the ruby returning to the dandy's palm, the diver lifting off from her imminent belly flop and returning to her tense position on the diving board. Let time expand, and speed increases; let it contract, and speed decreases, conveying in the case in which no time has elapsed the familiar fact that an object at rest has no speed whatsoever.

Whatever is to fall now begins to fall. The long elegant hand of the universal clock moves fluidly. And *now* one unit of time has elapsed: it is gone for good. And *at that very moment,* when the spider hand stands at 1, how fast is something falling? The request is for a concept of speed adequate to the very moment when the universal clock has tolled a single unit of time, its hand pointing inexorably to 1.

Galileo's law tells us how *far* an object has fallen in one unit of time. It has gone sixteen feet and so reached a position eighty-four feet from the ground. Neither the calculus, nor anything else, can freeze time in its frame. The hand of the universal clock having reached 1 moves next from 1 to 2. In this, the next second of time, the object drops a further forty-eight feet to reach a position thirty-six feet above the ground.

And here is what the reader must remember—that although the universal clock keeps ticking, it is 1 that is the pivot, the time that must be assigned a speed. Δt is therefore the difference between 1 and some later time on the universal clock, and since the mathematician next attends to the clock as it reaches 2, that difference is one unit of time, a simple second if the universal clock has been set to sound the seconds. Correspondingly, ΔP is the distance covered by a falling object between the first and second moment of its descent. That distance is expressed as the difference between its position after one second has been used up and its position after two seconds have come and gone—forty-eight feet in the present case, the difference between its position at eighty-four feet (one second is going) and its position at thirty-six feet (two seconds have

Clock and Tower

Galileo's law says that distance is a function $P(t) = 16t^2$ of time. In order to accommodate the height of objects dropped above the surface of the earth, the relationship is expressed as $P(t) = -16t^2 + 100$. The tower is one hundred feet high, and the object is falling *downward,* so that the 16 is negative in order to indicate the position the object has reached on a coordinate axis. If the value of t is 1—the hand of the clock has reached the first unit—then Galileo's law says that $P(1) = -16t^2 + 100$. And this is 84.

If the value of t is 2, Galileo's law says that $P(2)$ is $-16 \times 2^2 + 100$. And this is 36.

The difference *between* these two positions is ΔP, and this is $36 - 84$ or -48. When differences are recorded as distances, negative signs are displaced. The distance covered is 48 feet (if feet are chosen as the units of measurement).

The difference between these two times is Δt and this is $2 - 1$ or 1.

The ratio of *this* distance to *this* time is $\Delta P/\Delta t$ or 48/1. And this is the *average* speed the object has fallen over the interval between one and two units of time. Precisely the same calculations are continued with each new time.

If Δt is 0.5, $\Delta P/\Delta t$ is 40 feet per second; if 0.1, then 33.6 feet per second; if 0.01, then 32.16 feet per second; if 0.001, then 32.016 feet per second; and if 0.0001, then 32.0016 feet per second.

---

gone).[1] The average speed achieved between the first and second second marked by the universal clock is thus the ratio of the difference between these positions and the time that has elapsed: $\Delta P/\Delta t$ is 48/1, an *average speed* of forty-eight feet per second.

Time now undergoes its obligatory contraction, the difference in

---

[1] The elegant shorthand afforded by this notation should not obscure the fact that position is, position *remains*, a function of time, so that starting at the time $t = 1$, ΔP has the force of the expression $P(1 + \Delta t) - P(1)$.

time between the moment when the hands of the universal clock stand at 1 and successive moments shrinking. To speak in this way of time taking on different values is to appeal to *shorter and shorter* intervals of time. In the first instance, the hand of the universal clock moves from 1 to 2. In the second, from 1 to the position halfway between 1 and 2 at 1.5. In the third, the hand crawls only one-tenth of the way across a second, coming to rest at 1.1; and each refinement, I must take pains to stress, represents not so much a literal contraction of time, but a contraction of the mathematician's *attention* to smaller and smaller intervals of time, the contraction made possible by the original, all-controlling act by which the real world, including time's forward flow, came to be represented by the real numbers.

When the time elapsed stands at a full second, an object's speed stands at forty-eight feet per second. Each successive contracted interval of time is represented by a perfectly determinate, an entirely robust and familiar real number, things becoming small, but never infinitely small, and to each contracted interval of time, there corresponds an average speed, the rosy ratio of distance covered, however diminished, to time elapsed, however tiny. Starting at the moment when time stands at one unit, the contraction of time over the succeeding intervals of 1, 0.5, 0.1, 0.001, 0.0001, and finally, 0.00001 parts of a second yields the following average speeds, which for emphasis the reader must imagine chanted along some reverberating corridor in which memory, myth, magic, and mathematics are mingled: 48 feet per second, 40 feet per second, 33.6 feet per second, 32.16 feet per second, 32.016 feet per second, 32.0000 feet per second, whereupon the deep bass voice announcing these average speeds triumphantly intones: *Ladies and gentlemen, we have reached our limiting velocity at 32 feet per second,* the conductor's voice reminding reader and rider alike that it is the concept of a limit that has been loitering ostentatiously in the background all along, dying to be of use. As time contracts, the *limit* of the ratio of distance to time is the number 32:

$$\lim_{\Delta t \to 0} \frac{\Delta P}{\Delta t} = 32.$$

As time contracts? Δt is shrinking. It is approaching 0. *Approaching* 0. Differences in times are getting smaller and smaller. Average speeds are getting *closer and closer* to the number 32? The ratio $\Delta P/\Delta t$ is approaching 32 as Δt approaches 0. *Approaching?* The differences between $\Delta P/\Delta t$ and 32 are becoming ever smaller, vanishing under the hard white light of a limit.[2]

The symbols invite a mental motion in which two processes are at work. The first is one in which with a brilliant series of stroboscopic lights, time is flash-frozen at successive instants, its stream rudely halted, like one of those detective movies in which the grizzled veteran, examining for the one-hundredth time the accidental video of the crime scene, lights a cigarette and suddenly says *hold it,* asking that the action be frozen at a frame, and then, *roll that back,* freezing the action at an earlier frame, until he discovers the subtle clue that no one noticed: the gold earring on the carpet, the dog hairs in the hairbrush, the corpse showing alarming signs of life after death.

And the second motion is one in which at each frozen flash point a calculation is undertaken, a set of numbers juggled, almost as if that grizzled old detective were to say, as these characters always seem to say to their leggy blonde assistants: *You see that, Doreen?*

Doreen shakes her trim poodle cut in admiring perplexity.

Whatever it is that the grizzled old detective wishes Doreen to see, *I* want *you* to skip those hairs, the errant earring, that lively corpse, and compute only the average speeds for each interval of time, letting the intervals shrink and shrink until with a miraculous coordination of con-

---

[2] This formulation, which remains within the confines of a somewhat slangy brand of English, may, by means of the precise definition of a limit entombed within the appendix to chapter 12, be made as precise as the definition of a limit itself, the requisite precision coming about as the definition of a limit is specialized to the case of speed. To say that 32 is that instantaneous speed an object achieves at the moment $t = 1$ is to say that for any real number $\epsilon > 0$, there is a real number δ such that if $\Delta t < \delta$ then $\Delta P/\Delta t < \epsilon$. ($\Delta t$ and ΔP stand for distances and so are positive: no absolute values are required.) The essential idea to the concept of instantaneous speed is that of average speeds converging to a limit; but once that idea is accepted, the precise definition of a limit handles any further details. But at a price. When expressed within the standard formalism of mathematics the definition becomes very difficult to grasp and the brilliant simplicity of the underlying idea is very often lost, very often for good.

cepts *you* see the numbers that result approaching their limit, tapering downward toward 32 as the elapsed time tapers downward toward 0.

Enter Thus Speed

At one side of the universe the request has arisen for a number representing instantaneous speed, a number that might be correlated with time, a number constituting thus one half of a function; at the other side of that selfsame universe, a number has just been evoked. And, of course, surveying the whole scene in hindsight anyone at all can spot the romantically inevitable next step. The mathematician, who is the source and center of all these vagrant desires, compresses the edges of thought so that they come to touch. The longed-for number representing instantaneous speed and the limit toward which a sequence of average speeds is tending are declared in a superb, a redemptive, imaginative act to be one and the same:

$$\text{INSTANTANEOUS SPEED} = \lim_{\Delta t \to 0} \frac{\Delta P}{\Delta t}$$

Numbers dance now to an obviously motivating melody. An object falling for one second from a tower one hundred feet in height falls sixteen feet to reach a position eighty-four feet above the ground. Falling for two seconds, it falls a distance of sixty-four feet to a position thirty-six feet above the ground. Its average speed between one and two seconds of elapsed time is forty-eight feet per second, but the limit of average speeds as time contracts toward the first second is thirty-two feet per second. It is *this* speed that the mathematician confidently assigns to the time $t = 1$ as the hands of the universal clock move toward and just touch the number 1.

Pointing toward the classroom blackboard, where I have drawn a raggedy tower and a spastically falling object, the object's falling motion indicated by what I can control of the cartoonist's art, the whole thing an insultingly insufficient representation of the glowing panorama that *I* always see when I think of speed, I tap the tower meaningfully with the tip of my pen.

"What about it? The stone's instantaneous velocity?"

A student I have come to classify as Hafez the Intelligent coughs. He is a young man from the Middle East. I am spared the full force of his ferocious intelligence only by virtue of his inadequate English.

"This I do not understand," he says flatly.

"Now, Hafez," I say.

"If speed is change in position," he says in his guttural brand of English, "and change in position, it take time."—Hafez the Intelligent adjusts his delivery so that his concluding remarks will have the greatest possible force—"how can stone fall in *no* time?"

Hafez turns from me to the class, expecting his companions in crime to assist or at least applaud his efforts to derail the calculus before ever it gets started. Just then, the bell in the hallway tolls the hour. Hafez the Intelligent scuttles down the corridor of memory, leaving his argument behind him.

But Hafez, Hafez, Hafez, wherever you are, you were right. The derivative is an artifact, the first of the great concepts of modern science that fails conspicuously to correspond to anything in real life. In order to express speed as a function of time, the mathematician is prepared to sacrifice common sense, he is prepared to sacrifice the intuitive definition of speed, in plain fact, he is prepared to sacrifice everything.

Going from average to instantaneous speed is a passage from dry country to wet, the thing done slowly at first, as successive average speeds come up for inspection, and then with a whoosh, the limit in sight, the heat finally broken and gone. The function $P(t)$ is *differentiable* at t, and the limit itself is the *derivative of P at t*, a point of time, a precise moment. Note the contingencies. That limit must *exist* if the derivative is to be defined; it may not. But at the very moment of time that the derivative is evaluated, time disappears, since ΔT shrinks to 0. The limit toward which things are tending is never actually reached by the ratio of average speeds, no matter how far extended. The analysis of the behavior of a function at a point is accomplished only by the analysis of its behavior in a *neighborhood* of that point.

And finally, there are the obvious abbreviations, the *derivative of P at a point c* going over to

$$\frac{dP(c)}{dt},$$

this last very close to the notation that Leibnitz introduced.

The definition of the derivative has so far achieved a pairing between *one* time and *one* number. Like continuity, differentiability is a concept defined *at a point*, a fact that may explain its delicacy. Still, the derivative came into being as the result of a long-standing demand that speed be associated with time by a *function;* instantaneous speed has made its appearance at a given point, but has not, an eternally complaining voice might ask, the original demand for a function gone unmet?

Not at all. The process by which the derivative of *P* was constructed for a given time—1, as it happens—may be repeated again and again for a variety of points and hence a variety of times. Differentiable at one point, the function *P* may be differentiable elsewhere. It is the process of taking limits at *each* of these points that defines the requisite function.

Seduced by Speed

The image that I prefer is that of a thousand sturdy wooden doors which when opened lead to stony cellars or dingy courtyards, wash flapping on the line, or to absolutely nothing whatsoever. The calculus is no less a door; but at the definition of the derivative, against all expectation, it is one that swings open and there beyond are perfumed hanging gardens, bears riding tricycles, dancing maidens with almond eyes, jade palaces. The gardens, bears, dancing girls, and palaces are easy to miss amidst the gray difficulties of the definition; but the extraordinary thing is that the ambition to express speed as a function of time is attained by means of the concept of a limit, so that even now, with only half the concepts of the calculus in place, the sense is strong that the demands made by this discipline, this discipline meets in a way not familiar from any other study.

In a sense, of course, the definition of a derivative has overshot its mark, achieving a success greater and more considerable than anyone intended; created to account for instantaneous speed, the derivative goes *beyond* speed to acquire an identity as a separate idea. It is the fate of instantaneous speed to pass directly into the life of physics, where it becomes the cornerstone of mechanics, the rich mediational link between the acceleration a moving body undergoes and the position it attains, but differentiability betokens a concept richer than speed, and this is the source of its seductiveness. Speed is tied directly to the behavior of moving bodies. The derivative of a function expresses the more general concept of a *rate of change;* and the functions that are differentiable give voice to a notion of *change* more general than change in position. A function is a moving witness to change in *any* form, one thing—its arguments—changing against a change in another thing—its values; if we instinctively understand these changes to be changes in position and changes in time, that is because we instinctively look to a familiar world in order to bring illumination to general ideas. Not a bad policy, of course, but beyond any of its instantiations, the concept of differentiability has a level of generality, of spaciousness, denied to any of the concepts of physics or mechanics; it is the abstract record of *any* form of behavior, and hence any form of change, that may with profit be analyzed at a point, locally.

Newton and Leibnitz alike recognized speed as an essential concept, Newton especially requiring instantaneous velocity in order to carry out his work in mechanics; but what Newton and Leibnitz saw in the harsh but thoroughly unfocused glare of their own genius, they never clearly or completely understood, Newton writing of fluxions and fluents, ghastly mathematical ghosts wandering fitfully and moaning even after two hundred and fifty years, Leibnitz inventing a supremely flexible but deeply distorting notation. With neither of them able cleanly to break from infinitesimals or ever to say clearly what the derivative of a real-valued function *was,* they immured the calculus in incoherence. The concept of a limit purges the calculus of its logical impurities; but the concept does more. In making possible the definition of the derivative, the concept of a limit unifies in a fragile and unlikely synthesis two diverse aspects of experience, the discrete and the continuous.

It is discreteness that is most apparent. We are all separated from one another and from the world by the fact that we inhabit a body that is distinct from other bodies; segregated by the surface of our skins, we occupy a single unit moving amidst other single units, and this aspect of our experience is represented by the real numbers; for whatever the numbers, no matter how close they are to one another, the gap between any two of them is absolute and unbridgeable. But there is also the contrary experience of the world revealed by consciousness itself, as when in its forward motion experience seems to surge, the difference between one state and the next or between one time and the next becoming no difference at all, aspects, instead, of a continuum, or there are those queer moments when consciousness appears to melt moltenly into the world, to swarm beyond itself, as in the mystical meditative states when the commonsensical, the numerical, division between what I am and what there is blurs and then breaks entirely as the human soul, like water reaching finally the open ocean, becomes one with the ineffable. This experience finds its own reflection in physical processes, by the sensuous and unified way, for example, in which the surface of a river changes, where no one thing changes because on that surface there are no things, but where nonetheless the whole undergoes a transformation, the river reflected in its parts and its parts reflected in the river.

It is the genius of the calculus to reconcile in the definition of a limit and the definition of the derivative these conflicting aspects of experience. A real-valued function is invoked: it represents a continuous process, the change in the position of a moving object. The limit of the function at a point is a number, and it is discrete; but it embodies, it expresses because it is *their* limit, infinitely many numbers that in converging represent the continuous way in which the object is changing its position. So, too, the derivative of the function, which *expresses because it is their limit,* infinitely many numbers that in converging represent the continuous way in which the object is changing its speed. No escape from a world of numbers is ever made directly. The yielding and continuous world intimated by our experiences is nonetheless expressed by the driest and most austere of mathematical notations.

Speed
from Symbols

The tentative conclusion is that 32 is the limit toward which a series of numbers is tending. I say *tentative* if only because the case is purely circumstantial. Nothing has yet been demonstrated. A more elaborate argument now follows. The universal clock has reached the number 1. The demand is for a single number to represent speed *at this very time*. The value of the variable t is simply 1.

To derive the speed at this time, the mathematician considers temporal intervals Δt from 1 to other times and allows them to contract. The *position* of an object falling for each of those temporal intervals Δt may thus be written as $P(1 + \Delta t)$—the position of the object when one unit of time and a little more has gone by, the little more signified by the elastic Δt. Average speed has an incarnation in a familiar formula, one adapted to the forthcoming contraction of time:

$$\text{AVERAGE SPEED} = \frac{P(1 + \Delta t) - P(1)}{\Delta t}$$

$P(1 + \Delta t) - P(1)$ is, of course, ΔP itself, rewritten in a way calculated to make algebraic manipulation more transparent.

When Δt is allowed to contract toward 0, what emerges is the definition of instantaneous speed:

$$\text{INSTANTANEOUS SPEED} = \lim_{\Delta t \to 0} \frac{P(1 + \Delta t) - P(1)}{\Delta t}.$$

But the position function $P(t)$ has a perfectly explicit meaning. Whatever the value of t, $P(t)$ signifies $-16t^2 + 100$. In particular, $P(1 + \Delta t)$ may be rewritten as $-16(1 + \Delta t)^2 + 100$. Here the mathematician treats Δt as if it were a single variable, something amenable to the arts of algebra. And high-school algebra reminds us that $(a + b)^2$ is simply $a^2 + 2ab + b^2$. It follows that $(1 + \Delta t)^2$ is $1 + 2\Delta t + \Delta t^2$. Undertaking its work at the point $1 + \Delta t$, the position function yields

$$P(1 + \Delta t) = -16(1 + 2\Delta t + \Delta t^2) + 100.$$

The value of the position function at time $t = 1$, is

$$P(1) = -16 + 100.$$

The *difference* between $P(1 + \Delta t)$ and $P(1)$ is

$$-16(1 + 2\Delta t + \Delta t^2) + 100 - (-16 + 100).$$

And the ratio of this distance to the time that has elapsed is

$$\frac{-16(1 + 2\Delta t + \Delta t^2) + 100 - (-16 + 100)}{\Delta t}.$$

By elementary algebra, this reduces to

$$\frac{-16 - 32\Delta t - 16\Delta t^2 + 100 + 16 - 100}{\Delta t},$$

whereupon again by elementary algebra, *this* reduces to

$$-32 - 16\Delta t.$$

But it is the *limit* in which we are interested, and as the Δt shrinks ever closer to 0, the product of 16 and Δt itself approaches 0. But -32 minus an ever smaller amount tends inexorably to just -32, which is 32 absolutely, when the minus sign drops away.

This is the instantaneous speed assigned by the mathematician in this ingenious game to the number 1.

Not quite a proof, purists will observe, but it is close enough so that only purists will scruple; in any event, the evidence derived from formal manipulations corroborates the empirical evidence generated by an informal look at the behavior of P with respect to a handful of arguments.

If instead of assessing instantaneous speed at the particular moment when $t = 1$, the mathematician assesses speed generally, the same calculation, with the variable t introduced in place of the number 1, shows that whatever the time t, the speed of a falling object is $32t$, the derivative of the position function emerging as a simple function of time.

chapter 17

The Dimpled Shoulder

ONCE I WAS INVITED TO DELIVER A LECTURE AT THE UNIVERSITY OF Vienna. Although nearly everyone in Vienna speaks some English, the German language covers the city like a cloud, the inflected language giving rise to an inflected life. German or English, I asked my host, a mathematician who prided himself on his cosmopolitan command of the language of James Fenimore Cooper. "Oh, English, of course," he said. "We are all very sophisticated here."

Very well, English.

At eight, the university had pretty much shut down. The lecture room had the empty, somewhat spooky feeling of a large space put to little use. My friends were there, of course, and a handful of mathematicians from the university, serious men, with soft white hair and delicate hands locked in suspension underneath their chins; and a poet I had met at a party, who wrote a kind of furious, indignant German— *Warum? Wie so? Nein! Darf Ich?*—in which stray snatches of this and that would be modulated by a good deal of bizarre punctuation.

I began my lecture and realized promptly from the vacant expressions on the faces in my audience that no one understood a word of what I was saying.

Afterward the poet came up to me, the froth of remonstration about to form on his lips.

"I know," I said, "better in German."

This little memory came back to me when I read somewhere that Cauchy himself had traveled to Prague in the mid-1830s, in order to accept a position as a tutor to the morose and uncommunicative offspring of an exiled Bourbon, and the image of this impatient brilliant figure volleying his bright brittle French at uncomprehending and stolid Czech servants and chambermaids proves not only irresistible as the human expression of linguistic frustration, but emblematic somehow of a double vision that occurs not only in life but in mathematics as well.

Instantaneous speed is the solid embodiment of a significant concept. It was Newton in the seventeenth century who saw the need for a concept such as instantaneous speed and who cleared his mind of the cant that there must be a close, or even an intelligible connection between what a concept means and how it is applied; but it was Cauchy who in the 1820s found the system of delicate adjustments and inequalities required to express instantaneous speed in essentially its modern form. In the intervening years, a whole series of brilliant mathematicians, including Euler and Lagrange, extended the process by which an inchoate stirring becomes in the end a precise mathematical definition.

But if the derivative of a real-valued function is the mathematician's answer to the question *how fast?*, it is also the answer to an apparently different question, one dealing with curves and curvature and soft voluptuous shapes; in moving from one question to the other, the mathematician passes from a hard-edged utilitarian world to one that is brush-soft, my English in Vienna like Cauchy's French in Prague, an alien instrument amidst all those softly exploding sibilants and fricatives; and one of the enormous pleasures of the calculus, indeed, of mathematics, is the permanent possibility of seeing or sensing just behind an otherwise familiar facade the lineaments of an entirely different, infinitely more enticing world, the two worlds, the familiar and the fabulous, *both* under the control of the same system of mathematical ideas.

A sturdy straight line takes off from the origin of a Cartesian coordinate system. For every unit that the line moves out along the t-axis, it moves one unit upward along the y-axis. The equation describing its behavior is $y = mt + b$, where m is the equation's slope and b the y-intercept of the equation.[1] This line ascends from the origin, and b is thus 0. Now m in the equation denotes the juxtaposition of the line's upward and outward movements and so constitutes a ratio that may be expressed entirely in coordinate terms:

$$m = \frac{y_2 - y_1}{t_2 - t_1}.$$

The correlative distances in numerator and denominator ($y_2 - y_1$ and $t_2 - t_1$) measure the base and side of a familiar plane figure, an ordinary triangle, and function thus as numbers attached to things—lines, in fact—that have an existence beyond number:

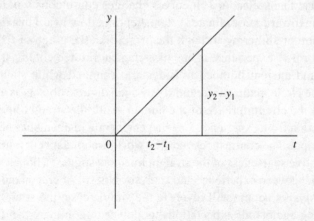

Slope of a straight line

---

[1] See chapter 4 for more detail.

The slope of this straight line is 1, and wherever it goes, that line, its address at a single point anywhere in space is all that is required completely to express its behavior. Three numbers in all, the one measuring the line's slope and the other two its address in coordinate terms, serve to indicate the line's permanent and unerasable relationship to the coordinate system. As in so much of mathematics, the simplest of examples is both rich and strange. The identity of a geometrical object has been characterized by its incidental attachment to a single point and a number measuring its angle of inclination, the analysis seeming to move *from* the larger geometrical world of straight lines *to* the less palpable but curiously more concrete world of numbers, so that once the requisite numbers are provided the feeling is strong, if unanalyzed (and perhaps *unanalyzable*), that like biological molecules the numbers contain the relevant information somehow within themselves, even the tritest and most ordinary of numbers capable by means of an enigmatic mechanism of embodying information sufficient to span the whole of space from one point of the everlasting to the other. But the slope of a straight line is also a strange and singular number: it is the same wherever the straight line goes. I say that this is singular because in the jostling congregation of lines embedded on the plane only straight lines have this property. The softest of soft curves changes continuously in its orientation toward a coordinate system, bending downward in a mournful droop or slithering through the origin like the graph of $f(t) = t^3$. The world of experience is one of arches and fluted columns, roadways and ancient humped hillsides and rounded white shoulders. Like the circle, realized irregularly by such diverse objects as a quarter and the circumference of the human skull, the straight line is an abstraction from experience, one that enters into mathematics because it is simple. The contrast between the world's abundant curvature and the relative sparseness of the straight line may suggest a hopeless distinction between experience and analysis. Not so. At crucial moments the analysis to come will revert to the simple case, the straight line retaining against all expectations its power to dominate the discussion, to inform, and to illuminate.

The function $f(t) = ct^2$ has made an appearance in the analysis of speed (where its values appear as $-16t^2$). Now let c decline drably to 1, $f(t)$ appearing in a new guise, like the house model at a fashion show, touchingly ready always to be on display. Its graph shows up on the now permanently blazing screen of a coordinate system as a *curve*, something that is in its heart of hearts a creature of change.

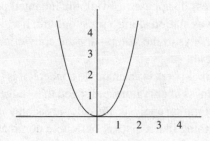

Face of $f(t) = t^2$

A black-eyed beetle moving sinuously upon this curve changes at every moment its orientation toward the coordinate axis. The distinction between what is taking place locally on the curve and the curve as a whole now assumes a new importance, a position of conceptual prominence. If there is a single number that characterizes the way in which a curve is changing throughout the *whole* of its expanse, it is not a number that the calculus provides. But it is natural to ask— it is a safer and a smaller question—whether *at any given point* there is a number that characterizes the way the curve is changing *there*. Such a number would embody and so express the degree to which a curve was changing *even as it changed,* and while it could not characterize curvature in the large, it might come separately to characterize each point in the curve, answering the larger question by the simple expedient of addressing the smaller question innumerably many times.

It is the concept of a slope that is wanted, a number that embodies information about attitude and change, but in considering curves, an old affliction returns to haunt the scene. The slope of a straight line is a single unchanging number, but like average speed it is a number derived from an assessment of two points. A straight line having the same slope at each of its points, this is a stricture without significance. But *curvature at a point* would seem to tremble on the same margin of incoherence as speed at an instant. The moving finger meanders over the rounded shoulder and stops, creating a pressure dimple in the molded flesh, but the hand having stopped, the sense of curvature conveyed by the caress disappears as well, the indented point being simply what it is, a way station along a sensuous arc; it is the *whole* of that shoulder that conveys to the voluptuary the conviction that he is getting anywhere at all.

Curvature, like speed, is an anarchic concept; it is the suspiciously similar waywardness of curvature and speed that suggests to the mathematician their common analysis, the calculus achieving its first act of mastery in something like a display of double domination. The curve that corresponds to the equation $y = t^2$, and so expresses the graph of the function $f(t) = t^2$, is a gentle parabola. An insect in motion on the curve, that beetle, say, descends toward the origin, leaving an iridescent trail in its delicate wake, touches the origin at 0 (since 0 when multiplied by itself is still 0), and thus invigorated, ascends through the curve. At $t = 1$, $f(t)$ is 1 as well, and so < 1, 1> marks a point on the curve, the soft dimple on the shoulder, the pair of numbers functioning as that point's coordinate. It is here that the beetle stops or the lover pauses, and here that the demand arises that the behavior of the curve be subordinated to a number.

The lesson of love is not to linger too long at a point, the same lesson, curiously enough, taught by the calculus. At $t = 2$, $f(t) = 4$, and the pair <2, 4> designates another point on the curve, a place further out on the shoulder. Between any two points on the curve, Euclidean geometry affirms, a straight line can be drawn, and this is one of those simple, inexpungeably vital *facts* upon which the calculus in the end depends. A straight line connecting two points on a curve is known as a *secant* line, *secant* from the Latin *secare*, to cut. A *straight* line, note, and thus a line with a well-defined slope. Well-defined? Defined how? Defined in the usual way, with the same triangle making a second no-nonsense appearance. Defined thus: as the ratio of $y_2 - y_1$ to $t_2 - t_1$, which in the case of <1, 1> and <2, 4> comes to $(4 - 1)/(2 - 1)$ or 3.

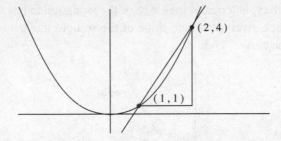

Secant lines may be drawn between any two points on the curve and form a delicate lattice of straight lines spanning the curve. The interposition of straight lines between points on the curve represents the reappearance of the elemental amidst the delicate tracery of the curve itself; but beyond the aesthetically pleasing contrast between the curve and the lattice of its straight lines is the rewarding fact that *each* secant line is satisfyingly straight and so has a well-defined slope, a number that defines its relationship to the coordinate axis itself, a flamboyant emblem of its position in space.

The secant lines touching a curve at two points are designates of a weak concept, one measuring the degree to which the curve itself is changing between points. In this regard, numbers representing their slopes play a role in memory as representatives of an average, a rough and ready assessment of curvature. Curvature under these aspects is curvature taken at two times and thus at two points. Between those points the curve itself can meander, turn in on itself, flatten out, or otherwise behave indecorously. The concept that emerges fails to fall under the sway of any function and so like average speed remains outside the great wheel of concepts the calculus is prepared to countenance.

A straight line *just touching* the curve at the single point <1, 1> where curvature needs to be assessed is different. It is here that, speaking loosely, curvature has been compressed, and it is here that the mathematician conceives the great dreamy idea of assigning the slope of a *tangent* line to a point on a curve, making the curve's soft roundness subordinate to a number, the lovely word *tangent* suggesting tangerines, tangelos, tangos, and the Latin *tangere,* to *touch.* The redemptive idea that will in the end assign a number to a point involves a shameless act of appropriation. The slope of a curve at a point is defined by proxy, in terms of the slope of the straight line just touching the curve at a given point, the slope of the straight line being passed along to the curve.

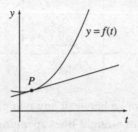

The tangent line to the graph of f at P

And yet this seems a conceptual gesture too simple to be useful. The tangent line has a spectral identity acquired from the fact that it touches the curve at the point <1, 1>, but the identification, such that it is, is tragically undetermined. Any number of distinct lines may touch a curve at a given point, and efforts to specify the line by insisting that it *just touch* the curve or touch the curve at *only* a single point are, as even Hafez the Intelligent acknowledges, his unshaven chin jerking in the air, doomed to disappoint.

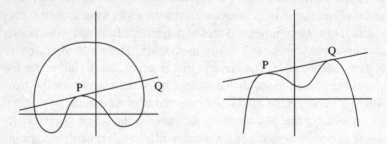

Tangents touching curves at many points

Several lines of thought and desire are about to meet, but before they do it is worth remembering how often liabilities may be converted into assets. However undetermined a tangent line may be, it appears in this discussion with one part of its identity already fixed firmly in place: it is a line that meets a curve *at* a given point. Completely fixing its identity is a matter merely of coming up with another number.

The mathematician intent on donating the line's slope to a curve, and the skeptic bemused by the fact that lacking a slope the tangent line is mathematically underdetermined, may be satisfied alike by a procedure that *assigns* a slope to the tangent line, the neutral idea of an assignment conveying, I think, the odd commingling of discovery and definition that is involved in any mathematical advance.

By the *retraction* of the secant line I mean the process of drawing the line back along the curve so that the distance between the points it spans grows ever shorter, the image conveyed by the operation of a fan closing.

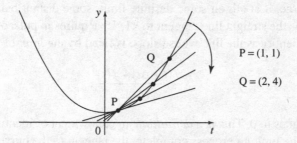

Retraction is something seen, but what is crucial beyond the snapshots of the fan closing is the *slopes* of the variously retracted straight lines, and these are expressed by a familiar formula:

$$m = \frac{f(t + \Delta t) - f(t)}{\Delta t},$$

various values of m arising as various smaller and smaller values of Δt come under consideration.

A few quick calculations. Allowing his eye to move down the curve and toward the point of interest, which is <1, 1>, the mathematician considers shorter and shorter secant lines. The first extends itself between the points <1, 1> and <2, 4>. The initial point at <1, 1> stays the same, fixed as a pivot, but subsequent secant lines are extended from there to the points <1.5, 2.25>, <1.3, 1.69>, <1.1, 1.21>, <1.01, 1.02>, and <1.001, 1.002>.[2] Associated with *each* of these

---

[2] Each pair of numbers is derived in the same way. The first number is a little past 1, and the second is derived by squaring the first. This is what the function *f* does.

straight lines is a slope, a number expressing one of the line's intrinsic properties. As the secant lines retract, growing shorter over these intervals, their associated slopes form the following sequence:

$$3, \ 2.5, \ 2.3, \ 2.1, \ 2.01, \ 2.0010, \ \ldots,$$

one that by now should prompt the suspicion that these numbers are converging toward a limit at the number 2.

As, indeed, they are.[3]

Retraction is a limiting operation, both visually, as the secant lines grow shorter, and analytically, as the slopes of those straight lines converge, if at all, on some definite limit, some distant ball of light, even as the straight line tangent to <1, 1> acquires in pulsed bursts its own identity as the line whose slope is fixed by the *limit* of

$$\frac{f(t + \Delta t) - f(t)}{\Delta t}$$

as Δt goes to 0. This is a *definition,* an endowment of identity.

The limiting process complete, the tangent to the curve emerges as a line with a firm mathematical identity, its place in the scheme of things fixed finally by two facts: that it touches the curve at $t = 1$, and so has the coordinates <1, 1>, and that its slope is 2.

The formula denoting the slope of a secant line admits of a notational reformation, a change of symbols, one of the ways in which a spark of insight may be coaxed from a symbolic shuffle. In what has just been given, the ratio of $y_2 - y_1$ to $t_2 - t_1$ has been expressed in terms of coordinates; it may be expressed, and expressed to the same effect, in the notation of the original function f itself, whence

$$\frac{f(t_2) - f(t_1)}{t_2 - t_1},$$

or what comes to the same thing

$$\frac{\Delta f}{\Delta t}.$$

---

[3] For a demonstration, see the appendix, p. 185.

This formula is plainly indistinguishable from the formula for average speed. Indistinguishable? Except for the fact that f is named f and not P, it is one and the same. At the limit, average speed becomes instantaneous, the passage expressed by the formula

$$\lim_{\Delta t \to 0} \frac{\Delta P}{\Delta t},$$

but at *their* limit, the slopes of retracting secant lines become the slope of the tangent line, the passage expressed by precisely the same formula

$$\lim_{\Delta t \to 0} \frac{\Delta f}{\Delta t}.$$

Various concepts having been conjoined, the question of how a curve curves is given an elegant, simple answer. Curvature is assessed at a point by reference to the slope of a line tangent to the curve at that point, the curve acquiring its slope at second hand, it is true, but acquiring nonetheless a slope and so a number embodying and then expressing its curvature. An intricate exercise in abstract thought is on display in these brief remarks and the picture with which they are correlated, an effort the more remarkable because it appears to have been made with ease, each step following naturally from the one before, each step yet representing an effortless enlargement of the light. Speed is a concept whose origins lie with the experiences of the human body and so corresponds in its most primitive form to something *felt*. The analytic exercise in which average speed is expressed by a number marks the imprint of an arithmetic calculus upon these otherwise turbulent and unstable experiences. The discovery that *this very number* serves as well to measure the slope of a secant line endows the formula by which it is computed with a rich multiplicity of meaning so that in passing from a consideration of speed to a consideration of curvature, the mathematician has the impression that only one shimmering, multifaceted concept has been deployed and not two. In the same fashion, instantaneous speed and curvature at a point are represented by one and the same mathematical formula, one defining the derivative of a real-valued function, and this is a second remarkable but expected fact, almost as if two widely separated explosions were seen after the Zen master's salutary slap to be eruptions from a single, hotly glowing central sun. Speed appears as an aspect of curvature, the essence of instantaneous speed suddenly *visible,* the curve of the position function expressing the now discernible meaning of the concept; but curvature also appears as an aspect of speed, the essence of curvature undergoing an analytic compression so that if speed may now be seen, curvature may now be *computed.*

The derivative of a real-valued function emerges from these considerations as a masked concept, one that on some occasions appears as speed, on others as curvature; but however different the manifestations, the sense should be strong that some central conceptual substance in the very nature of the derivative has stayed the same so that this variety and multiplicity of meaning in mathematics is evidence that there is unity and identity and indivisibility in nature.

The Devil's Work

The calculus is both a great theoretical achievement *and* a unique set of computational tools, a collection of algorithms; and for more than three hundred years, it was the existence of the algorithms, the techniques of thought, that made the application of the calculus possible. The function $f(t) = t^2$ offers an example. At the point <1, 1>, where $f(1) = 1$, the derivative of f is the number 2. The derivative of a function is an intensely local object, a number matched to a point. The procedure by which 2 was assessed as the derivative of f was pretty much a matter of retracting a number of secant lines, computing their slopes, and then *guessing* at the limit toward which those slopes were tending. The argument may be clinched algebraically by an appeal to the definition of a limit.

The idea is to assess the behavior of

$$\frac{f(t + \Delta t) - f(t)}{\Delta t}$$

at the limit as $\Delta t \to 0$, *whatever* the time and so whatever the value of t.

The trick is to allow the functions to do their accustomed work. Remember thus that f acts to square things. With $t + \Delta t$ and t both squared, the result is

$$\frac{t^2 + 2t\Delta t + \Delta t^2 - t^2}{\Delta t}.$$

But when t^2 and $-t^2$ are allowed to cancel, and the numerator divided by Δt, the result is simply

$$2t + \Delta t.$$

As Δt grows ever smaller, its contribution to the sum of $2t + \Delta t$ becomes vanishingly small, so that the limit of $2t + \Delta t$ is simply $2t$ itself.

An effort of will is required to re-create the wonder of this little argument, a rolling back of time by three hundred years. Let calculating machines and calculus textbooks, computer programs, and the Internet be blotted from memory. The problem before us is to compute the derivative of $f(t) = t^2$ at any place the function f happens to land. It is a wearisome prospect, the argument of the text repeated over and over again; yet in the darkness of this deplorable work, the simple discovery that the derivative of a function whose form is t^2 is $2t$ lights up the scene like a thousand rockets. Limits need no longer be computed at all. At $t = 9$, the derivative of f is $2t$ or 18; at $t = 234$, the derivative of f is $2t$ or 468.

The fact that an algorithm exists by which the derivative of $f(t)$ may be computed does not at all cancel out the local color or character of the derivative. It simply means that at each point there is a procedure available for calculating derivatives.

And, indeed, there is a comparable procedure available for each of the elementary functions.

The derivative of the function $f(t) = at$ is a.

The derivative of the general power function $f(t) = t^a$ is ax^{a-1}, a result already familiar from the case in which $a = 2$.

The derivative of the function $f(t) = \ldots$, but really what is important is not so much the specific results as the fact that the results are forthcoming; and for this a list suffices.

The elementary functions form a collection of concepts in which things tend to work out well. The derivative of every elementary function is again elementary. Beyond specific results, there are a few rules controlling differentiation in a few pat situations. The derivative of a sum $d(f + g)/dt$ is, for example, the sum of their derivatives $df/dt + dg/dt$. There are other rules coordinating the multiplication and division of functions, and there is the *chain rule*. It is the chain rule that covers the case in which a function takes a function as its argument and so suggests one of those Sumerian creation myths in which Enog begat himself.

Derivatives of Some Elementary Functions

What follows is a collection of rules or recipes, each created in the following manner. An elementary function is given, and a way specified to determine its derivative by means of another function. The text already provides an example in the case of $f(t) = t^2$. This is the function. The function indicating its derivative is in turn 2t or, if the function is to be named explicitly, $g(t) = 2t$. Another way to name $g(t)$ is by means of Leibnitz's notation: $df/dt = g(t) = 2t$. The notation is confusing, if only because unfamiliar, but it should afford a sense of the fantastic flexibility of functional notation—indeed, the flexibility of the concept of a function itself.

---

1. $da/dt = 0$. Here a is a constant. The function $f(t) = a$ plainly is going nowhere; its graph is a straight line parallel to the t-axis, and so its slope is 0.

2. $dat/dt = a$. The function $f(t) = at$ describes a straight line; its slope is a itself.

3. $dt^a/dt = at^{a-1}$. The function $f(t) = t^a$ represents the general power function; the fact that the derivative of $f(t) = t^2$ is $2t$ designates a special case of the general rule, an example of things working out as one would expect.

4. $d\sin t/dt = \cos t$. Again as one might expect, the derivatives of the **sine** and **cosine** functions are promiscuously intertwined.

5. $d\cos t/dt = -\sin t$. Ditto.

6. $d \ln t/dt = 1/t$. The derivative of the natural logarithm $f(t) = \ln t$ is $1/t$.

7. $de^t/dt = e^t$. The exponential function is its own derivative, a bizarre and useful fact. No other function has this property.

How does this work? Begin with an ordinary function $g(t)$ and suppose that $g(t)$ is the argument of a still grander function $F(g(t))$.

You can do that?

Sure. Is $g(t)$ a function? Yup. It denotes a number, does it not? Yup again. So, then, make the value of $g(t)$ the argument of F.

That having been done, the chain rule covers the case in which the mathematician wishes to know how F varies *not* with $g(t)$ but with t itself. The answer the chain rule provides is that the derivative of F with respect to t is the product of the derivative of F with respect to $g(t)$ and the derivative of $g(t)$ with respect to t, and I have no sooner said this than I hear a susurrus of moans. The astonishing thing is that in Leibnitz's notation, the thing comes to be expressed in a concise, a marvelously simple, formula:

$$\frac{dF}{dt} = \frac{dF(g(t))}{dg(t)} \frac{dg(t)}{dt},$$

truly the Devil's own in its clarity, its capacity to compress information.

With the Devil's rule in hand, here is an example of the Devil's work. The function $F(t) = (t + t^2)^2$ determines its value according to the following calculation: Take t and then add it to the square of itself and then square *that*. What is the derivative of F at t? The problem is simpler than it looks. To begin with, $t + t^2$ is a function. Call it $g(t)$ on the understanding that $g(t) = t + t^2$.

But then $F(g(t)) = g(t)^2$ and the chain rule sings out that

$$\frac{dF}{dt} = \frac{dg(t)^2}{dg(t)} \frac{dg(t)}{dt},$$

$g(t)$ acting *both* as the argument of one function and as a function in its own right.

But the derivative of $g(t)^2$—that's just $2g(t)$. Remember $f(t) = t^2$? Same thing!

Next, what is the derivative of $g(t)$ itself?

Piece of cake. The function $g(t) = t + t^2$ is a sum. The derivative of a sum is the sum of its derivatives. So what are they, those derivatives?

Well, the derivative of t is 1.

"Where from this comes?" Hafez the Intelligent demands in his harsh, guttural English.

Good question, Hafez. What is the function we are considering? It is the function $f(t) = t$, now isn't it? And the derivative of this function is 1 because the derivative of at is a and here a is 1.

Now we *know* that the derivative of t^2 is $2t$. *You remember, we just did that.* Putting these pieces together, we see—*we?*—that the derivative of F at t is $2g(t)(1 + 2t)$. Substituting $t + t^2$ for $g(t)$—

Mr. Waldburger comes briefly to life in order to allow an expression of utter perplexity to cross his otherwise untroubled young features.

How come? We can substitute $(t + t^2)$ for $g(t)$ because we *defined* $g(t)$ as $t + t^2$.

Where was I? Yes, I remember now, the only one in class apparently who does. After making the substitution, $2g(t)(1 + 2t)$ comes to $2(t + t^2)$ $(1 + 2t)$.

Yes, the derivative of F at t is $2(t + t^2)(1 + 2t)$, and *yes,* this is the answer and *yes,* you do have to know the chain rule for the final, and *yes,* the steps that I have undertaken are moments in an ongoing miracle, despite the remembered ineptness of my own classroom presentation, and *yes,* the miraculous always takes precedence over the mundane, and takes precedence, too, over the difficulties involved in doing difficult work precisely. No calculator. No computer. No adding machine. Not even a word processor. No help from any quarter. No nothing save for the chain rule, and *yes,* the chain rule, and nothing else, has seen us—I might as well be fair to myself, it has seen *me—* through a complex deductive chain that without the chain rule would have remained hopelessly beyond my powers to execute.

As I said, the Devil's work. But what I should also say is that these complex and miraculous computational routines are now embedded in even handheld calculators, the ancient algorithms and tools of the trade destined to disappear as completely as Horner's method of root extraction and ultimately as completely as poor Horner himself.

chapter 18

Wrong Way Rolle

WHO? FOR MANY YEARS, MICHEL ROLLE FIGURED IN MY IMAGINATION as a poignant and inconspicuous blank. During the years that I taught the calculus, no one ever asked with a warm chuckle of human sympathy who he was or what he did. One day, provoked by a dream in which a well-dressed man wearing a homburg sees and utterly disregards a used copy of one of *my* books at a bookstall, I looked up *his* dates: 1652–1719. What else? He was born in the Auvergne, still to this day an especially rugged part of France, with ruined châteaux perched on the rounded hills, the grounds protected by moss-covered stone walls, deep gorges cutting through the wooded countryside. Sometime in his twenties, Rolle left his muddy village for Paris, where for a number of years he eked out a living as a scribe and a reckoner, someone, I gather, with enough elementary arithmetic to handle books and do numerical computations; but he seems to have been a man of genuine mathematical talent, his mathematical skills self-taught and by middle age, polished and considerable. His public solution of a difficult algebraic problem led to recognition from Colbert, the finance minister, and a pension or at least a stipend. He became a man of parts, a man of Paris. He published in the learned journals and entered the Academy in 1685; but what lends retrospectively to Rolle his perverse charm is the fact that in the last years of the seventeenth century, he participated in a public debate over the merits of the calculus, one of those boisterous affairs in which members allied with various factions at the Academy puff themselves up and heap abuse upon one another, Rolle himself arguing *against* the calculus, constructing endlessly ingenious arguments demonstrating that the concept of a limit was absurd and incoherent. As, indeed, it then was. On reading this I conceived an immediate sense of identification with poor Rolle, thinking of him irreverently as *Wrong Way Rolle,* an intellectual ancestor, the two of us a part of the great fraternity of mathematicians who myopically would be apt in any century to miss the main chance. And there the biographical notice stopped, leaving me to endow Rolle with a lavish head of black coarse hair, this suggesting vaguely Sicilian ancestry (and why not?), a low, ridged forehead, a fleshy nose with tense volutes, and sensuous lips.

And since this is all nonsense, fabricated from scratch, I might as well have him composing *his* theorem—Rolle's theorem—in an attic garret, his mistress blowing gently on the bedside candles in order to coax him to bed, *Michel, viens au lit,* come to bed, as Rolle, sitting on a wooden chair with a straw seat, says impatiently that he will be there in a minute; but this, too, is nonsense, this time on stilts, not the mistress, of course, but the theorem, for the identification of Rolle with his own theorem came about only in 1846, when an Italian mathematician by the name of Giusto Bellavitis, perhaps as the result of research among Rolle's papers, put together the long-dead Rolle and the theorem that now bears his name. Whether Rolle ever thought through the steps that I now attribute to him, I do not know.

Rolle's theorem is in the first place a theorem about functions and *so a theorem about processes represented by functions,* an affirmation among other things about the coordination of time and space. Let a function *f*, say, be given or imagined. Let the concepts of closed and open intervals be resurrected from memory's vault: The interval between two numbers *a* and *b* is *closed* if both *a* and *b* are among the numbers in the interval, and *open* otherwise, the two different kinds of intervals designated by [*a*, *b*] and (*a*, *b*).[1]

Now about that function. Suppose in the first place that *f* is continuous on the closed interval [*a*, *b*]. *Continuous* on the *interval*? Continuous at *every point on the interval*. Every point *including* the end points. And suppose, too, that *f* is differentiable on the open interval (*a*, *b*). Differentiable at *every* point *within* the open interval. And suppose, finally, that $f(a) = f(b) = 0$. These are specific constraints on the function *f*. They serve to pick out one class of functions, and so one class of processes, from all the others. The constraints deal with the two fundamental mathematical properties of continuity and differentiability; and they reveal the mathematician characteristically concentrating his attention, the world's delight emerging as it always does from the world's details.

The far wall of the garret now acquires the aspect of a Cartesian coordinate system, a crack parallel to the floor straightening itself out to become a coordinate axis. The contingencies that Rolle is contemplating resolve themselves into a simple curve rising from and then returning to the axis. Even though *I* am in possession of concepts clearer and more polished than any Rolle might have known, let me infuse Rolle with enough life to suggest the moment in which these jumbled contingencies shift and reveal the inexorable aspect of the obvious that Rolle has sensed but not yet seen. The curve rises from a crack in the wall, flows upward and rounds itself, and then descends back toward the plaster crack. Fixed at *a* and *b* by the fact that $f(a) = f(b) = 0$, the curve, acquiring a living aspect of its own, comes to resemble a stiff leather whip or even a sufficiently stiff length of starched rope, something held at either end so that by means of a rotation of one's wrists the rope's rounded hump seems to undulate through space, passing

---

[1] These delicate distinctions are *crucial*, and the reader inclined to mumble impatiently, "closed, open, big deal, what's the difference?" should appreciate in what follows that without [*a*, *b*] there is no maximum and without (*a*, *b*) no appropriate derivative.

from left to right and back again to left; the remarkable fact about this familiar image is that in some sense something about the curve is *controlled* by the fact that at *a* and *b* the function has the same value.

Just so, I am persuaded to interrupt, hurrying up the thoughts of my great predecessor. Whatever the ultimate shape of the curve (the various possibilities corresponding to undulations of that rope), if it returns to the place whence it arose at the axis itself, *the curve must change its direction* (and so its character), and this observation belongs to that peculiar world of affirmations (of which moral life is strangely replete) that are at once obvious and surprising simply because they are obvious.

Rolle's sloe-eyed mistress, her black hair spread over the muslin of their single pillow, has long since fallen asleep, a childish bubble forming on her full red lips; and as the moments pass in the seventeenth century and again in the twentieth, one thought engenders another, the movement of thoughts expressing an inferential chain so natural as to appear unforced, as breathing itself. The curve rises and then falls. *At the point where the curve changes its direction, a line tangent to the curve must be parallel to the coordinate axis:*

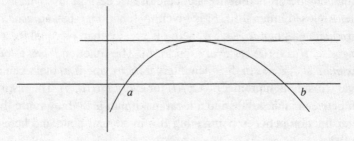

Once pointed out, this is something that anyone can see. Mathematics is far away. Humped over, the curve is touched by a tangent line, and *because* it is humped the tangent is parallel to the coordinate axis itself. But the language of tangents has an analogue in the calculus. Mathematics now comes roaring back with a rush. A tangent line parallel to the coordinate axis has a slope of 0, and I am now in the position of reminding Rolle what he himself might not have known—that in *these* circumstances, the crack in the wall straightening out for good to become a Cartesian coordinate system, there must be a number *c*

within (a, b) such that *at c*, the derivative of f is 0. A *number,* please note, one that induces a cold arithmetical shock into the warm geometrical landscape of this imagined domestic scene.

Bringing the various contingencies and their conclusions together yields a more formal statement of Rolle's theorem. I will handle the details myself. If a function f is continuous on an interval $[a, b]$ and differentiable on (a, b), then there is a number c in (a, b) such that the derivative of f at c is 0. I like to imagine that as I explain this modern version of his theorem to Rolle himself, a slowly dawning smile of sympathetic appreciation crosses his tough, creased face.

Rolle's theorem in its largest aspect says that change in the character of a continuous curve may be coordinated with a number, and although some unavoidable aspect of wooliness infects any informal statement of a mathematical theorem, there is the compensatory fact that seeing things in this large and general way illuminates the theorem's real achievement, which is to reveal a connection between the *property* of continuity, the *fact* of change, and the *existence* of a particular number.

And the proof? Childishly easy, as it turns out. Every now and then in mathematics, it is time for a couple of fast definitions. This is the time. A closed interval $[a, b]$ is given. A function f has an *absolute maximum* at a point, c, say, if there f is greater than or equal to anything else: $f(c) \geq f(t)$ for every t in $[a, b]$. The function f has a *local maximum* at c, if c is in the open interval (a, b), and if in that open interval $f(c)$ reigns supreme: $f(c) \geq f(t)$ for every t in (a, b). The distinction between an absolute and a local maximum is nothing more than the distinction between being a big fish in a small pond and being a big fish;

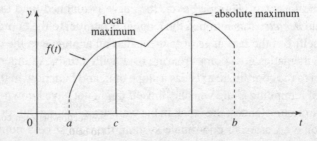

but the distinction also serves to emphasize an important point, one obscured by the symbolism, and that is the fact that being a mathematical big shot is often a *local* characteristic of a function, things being the greatest (or the smallest) by reference to the neighborhoods in which they find themselves. This is something, my students often remark ruefully when I put the point in just this way, that they knew without studying the calculus. But there is an additional point that hinges on the point just made. Quite often, conceptual connections run between local properties, the large absolute properties of a function too muscle-bound to be much use. And this, too, my students say, they knew all along, suggesting somehow a source of wisdom to which they have painful access.

And now a fact about the continuous functions. If f is continuous on a closed interval $[a, b]$—yes?[2]

There follows the memorable refrain: *Then f attains an absolute maximum at some point c within* $[a, b]$.

Such are the facts scattered like guest-stars on the grass. A little theorem by Fermat draws a connection among them. *If*, it affirms, that little theorem, a function f achieves a local maximum at c then at c the derivative of f is 0. The theorem makes sense, of course; I introduce it without proof.[3] A point that drives the derivative of a function f to 0 is a *critical* point; all other points are *regular*. A function answering to the assumptions of Fermat's theorem *has* a critical point, a place c where the derivative of f is 0.

Note the satisfyingly unexpected and stylish character of the theorem. A *local* maximum or minimum of a function is signaled by the function's *derivative*. And note, too, that the theorem goes beyond the mere assertion of a connection. Its conclusion issues in a *number*. The derivative goes to 0.

The facts left lying carelessly about amount almost to a proof of Rolle's theorem. The theorem says that with those *ifs* in place a number can be found; at that number, f answers to a sane and simple con-

---

[2] See chapter 15, p. 137.
[3] Indeed, in a local environment it reproduces the effect of Rolle's theorem itself. But Rolle's theorem is more general, drawing as it does a connection between continuity and the fact that somewhere the derivative of f is bound to be 0.

dition: its derivative is 0. Now, *f has* a maximum on [*a*, *b*] just because it is continuous. If the maximum is *f*(*a*) or *f*(*b*), the function as a whole is flat, and there is nothing to prove, *f* determining a straight line in space between *a* and *b*. The slope of this straight line is 0. Any point between *a* and *b* will do as the required *c*. The maximum must be somewhere in the open country (*a*, *b*). It must thus be a local maximum, and there, Fermat's little theorem says, the derivative of *f* is 0.[4] *Quod erat demonstrandum,* as Latinists like to say.

Rolle's theorem establishes a connection between continuity and differentiability. Continuity guarantees a maximum; differentiability delivers a number. Fermat's little theorem supplies the connection between concepts. And all this is almost trivial in its mathematical content. Yet Rolle's theorem quivers with significance. A ball is thrown into the air. At its extremities it *changes* its behavior. What was going up is now coming down, and *between* going up and coming down the thing must reach a point where it changes *from* going up *to* coming down. (Indeed, Rolle's theorem may be used to demonstrate that the velocity of a ball thrown upward must at some point be 0.) What seems to us a natural description reflects on a profoundly unconscious level the tendency that the calculus enforces to describe movement by means of continuous functions. The description imposes a certain characteristic structure on things moving in the real world. Something so simple as a ball in flight has acquired a tripartite aspect, a critical point lying between points marking its regular behavior. Like a great resolving telescope, the calculus reveals features of experience that would otherwise be obscure—that would otherwise be *invisible*.

In all this, two mathematical epiphanies are at work. The first occurs as the imagination records with a rush the movement from continuity to the existence of a number that takes place in passing *from* the fact that *f* is continuous in an interval *to* the fact that somewhere in that interval there is a point driving the derivative of *f* to 0. The second, experienced simultaneously, but with a larger and more leisurely wave of appreciative amplitude, occurs as the imagination begins to see in the structure of functions and their graphs, the hidden record of

---

[4] This is an argument, but not yet a proof. I have neglected the possibility that *f* might achieve a minimum on [*a*, *b*] instead of a maximum.

a second analytic world, that of derivatives and their properties, almost as if the scene revealed by Rolle's theorem were like one of those seventeenth-century Dutch canvases, in which a cleverly placed convex mirror reveals that beyond the domestic interior there is a distant landscape through a window, a faraway haystack, a row of drooping trees.

All this and more I can read from the wall that Rolle read.

chapter 19

The Mean Value Theorem

IN 1792, AS THE INK WAS DRYING ON THE AMERICAN CONSTITUTION AND tumbrels rattled through the streets of Paris, the great mathematician Joseph Louis Lagrange was rescued from melancholia by the attentions of a pretty young woman. All happy families, Tolstoy remarked more than half a century later, resemble one another, but each unhappy family is unhappy in its own way; and in this famous sentence the concepts of the calculus may be seen glimmering, the unhappy families marking singularities or isolated spots in the great manifold of human undertakings, the happy families regular points, alike in being regular, but distinct to the extent that they occupy different positions on the manifold. More than anything, it is the mean value theorem that enriches and justifies this vivid image of the singular and regular families, for it is, above all else, a theorem that provides the mathematician, and so by implication the novelist, with an instrument of analysis flexible enough to describe diversity within a family of similar characters.[1] What lends to these speculations their pathos is that Joseph Louis Lagrange, who first discerned and then demonstrated the mean value theorem, was himself an unhappy family all on his own, a one-man multitude of misery, suffering in his middle age from a sense of despondency prompted by a strange and terrible emptying of content from the world and his work, so that what he had accomplished seemed ignoble to him. Late in life, the great shadows lengthening everywhere, Lagrange married his mathematically illiterate and engagingly bubble-headed *acquisition,* and despite the natural expectations of his friends, who divided their concerns between indignation and envy, his marriage flourished; he became notably uxorious, devoted to his wife and she to him, and thus soothed and settled by the most palpable of pleasures, good food, the calm that regular domestic habits afford, the pleasures of a marriage bed, those sensual artifacts that stand in such remarkable contrast to the austerities of mathematical work, he undertook a revision of the *Mécanique analytique,* the great treatise on analytic mechanics that he had composed during his most productive years. In marrying happily Lagrange passed against all expectation from one side of Tolstoy's divide to the other.

Lagrange was born in 1736 and died in 1813; his was a vantage point that allowed him to see the great and historic events of the eighteenth and early nineteenth centuries explode like cannon shots. Although he spent the most productive years of his life in leisure as a

---

[1] The study of the way in which the arts mirror developments in mathematics is a fascinating one. Lipman Bers began a memorable lecture on non-Euclidean geometry by comparing the Declaration of Independence with Lincoln's Gettysburg Address. "We hold these truths to be self-evident," he rumbled in his Russian-accented English, quoting the Declaration, "that all men are created equal." And then he cited Lincoln's enormously cagy address: "*. . . conceived* in liberty and *dedicated* to the proposition that all men are created equal." The class of native-born Americans realized with a shock that Bers had discerned something we had never seen: Lincoln's frank unwillingness to commit himself unequivocally to the propositions he was citing. *Why the difference?* Bers would ask. The discovery of non-Euclidean geometries in the early part of the nineteenth century, he would rumble on, an interesting if playful and irresponsible answer.

member of the Berlin Academy, an improbable favorite of Frederick the Great, he was also the first professor of mathematics at the École Polytechnique, the greatest of Napoleon's *grandes écoles,* those democratic institutions devoted to the cultivation not so much of genius as of a remarkable form of competence, what the French themselves call *excellence de tout premier ordre.* It was from these schools that Napoleon withdrew the artillery officers and civil engineers who made possible his conquests, Cauchy himself acquiring a sense of his own powers first as a military engineer.

Like Euler and like so many members of the French intellectual aristocracy during the years from roughly 1740 to 1820, Lagrange appears in the history of mathematics in possession of an intellect which made possible great achievements, but by means of mental processes very unlike those evident among great mathematicians of the present era. He worked hard, his contemporaries agree, but he did not *labor,* the time devoted to research considerable only because he had so many things to say. He cultivated Newtonian mechanics, radically extending the scope of Newton's theories; he cultivated arithmetic and the theory of equations; he stated and then proved the mean value theorem in essentially its modern form. To this far-flung universe of concepts and concerns, Lagrange brought intellectual resources that I would describe as *cosmopolitan* were it not for the fact that in explaining what I meant I would, no doubt, lose myself in a thousand irrelevant detours and alleyways; yet like so many others of the French mathematicians, Lagrange does evince a sense of knowing his way around, and perhaps the secret of his singularity is something so simple as this—that in the latter years of the eighteenth century the sphere of mathematical knowledge was so much smaller than it is today that lavishly gifted individuals *could* circumnavigate the globe.

All Happy Families Are Alike

The mean value theorem is a statement about functions and their derivatives. Short of the fundamental theorem of the calculus itself, it is the most important theoretical affirmation of the calculus, one of the protean assertions of mathematics, a theorem with a thousand faces.

Suppose that a function f is continuous on $[a, b]$ and that it is differentiable on (a, b). So far nothing in the schedule of contingencies has changed. We are traveling in all the old familiar circles. But where Rolle's theorem goes so far, the mean value theorem goes further. Continuous in the closed interval $[a, b]$, the function f appears in any picture as a curve:

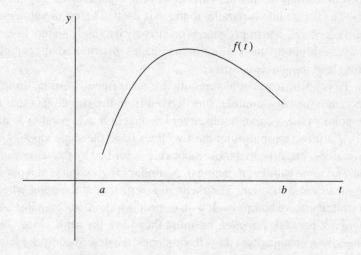

Now a *chord*, geometricians say, is a straight line **AB** connecting two points on a curve. (The secant lines of old are all of them chords.) The mean value theorem asserts in the lovely language of geometry that there exists on this curve a point whose tangent is parallel to the chord:

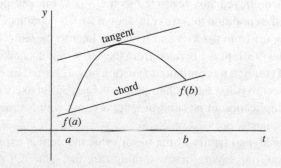

As pictures so often do, this picture compels belief. A curve, a point, a chord, and a line appear in space, connected only by the circumstance that if the curve meets certain analytic conditions the line is parallel to the chord. The family resemblance to Rolle's theorem is impossible to miss. This new affirmation is simply Rolle's original theorem applied to an arbitrary curve subtended by a chord. But Rolle's theorem restricts the mathematician's eye to a particular case, one in which a tangent line is parallel to the axis itself. The mean value theorem is general. The impression is one in which a single picture is suddenly multiplied, the various new images distributed dizzyingly throughout the whole of space.

Beyond the world of pictures, there is an analytic assertion. Rolle's theorem issues in a number, one derived from the simple fact that at the point where a curve humps over its tangent line is parallel to the axis. *Parallel,* meaning that the two lines have the same slope. *The same slope,* meaning that those slopes are described by the same number. *The same number,* meaning 0. A similar progression is at work in the mean value theorem. The theorem asserts that at the point where an arbitrary curve humps itself with respect to a chord, tangent line and chord are parallel. *Parallel,* meaning they have the same slope. *The same slope,* meaning that they, those slopes, are described by the same number; but where Rolle's theorem gives that number, the mean value theorem only affirms the identity. The mean value theorem does not— it *cannot*—go on to identify those slopes as 0. Tangent lines in the context the theorem contemplates need not be parallel to the axis. What the mean value theorem can affirm, and it is all that it can affirm, is simply that the slopes are the *same*. Whatever the circumstances, Rolle's theorem remains restricted to the t-axis, the end points of f bound by the condition that $f(a)$ must equal $f(b)$, the graph of the function taking off from the axis and returning later to the same axis. The mean value theorem is liberated from these restrictions and embraces *any* chord cutting *any* continuous function on *any* closed interval. The pleasant security of a particular number is lost, but in exchange there is the multiplication of possibilities, curves, and chords crowding the Cartesian plane.

The analytic identity of the mean value theorem is expressed by an equation, one saying that two things are the same. Of those two items, one is the slope of the chord **AB**, and that is an antecedently familiar quantity:

$$\frac{f(b)-f(a)}{b-a},$$

a familiar triangle making its inevitable pedagogical appearance:

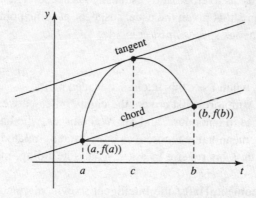

The other is the slope of a tangent line, which by definition is the derivative

$$\frac{df}{dt}$$

of the function f, the limit, recall, of retracting secant lines or the limit of average velocities as time dwindles toward 0. The mean value theorem affirms that *somewhere* the slope of the tangent line is the same as the slope of the chord. Somewhere, meaning *at some point.*

The equation to which the mean value theorem is committed is now in sight. The conditions first: if f is continuous on $[a, b]$ and differentiable on (a, b); and the fundamental claim second: then there exists a point c such that

$$\frac{df(c)}{dt} = \frac{f(b)-f(a)}{b-a}.$$

Such is the analytic information conveyed by the picture. It is possible to look long and hard at this equation without being gripped by a sense of its importance, without, in fact, quite understanding what it means. But the equation itself has an engaging and secondary interpretation that often convinces students all over again that its content *is* obvious. If f is taken as a measure of position—the *position* function of chapters past—then $f(b) - f(a)/(b - a)$ and $df(c)/dt$ have an acquired identity as average and instantaneous speed. The mean value theorem asserts that given the usual suspects, at some point c average and instantaneous speeds *must* coincide.

There is yet, when I say this in class, a leaden pall.

I say, "A motorcyclist covers the eighty miles between Barstow and Las Vegas in one hour," and Mr. Waldburger, who has until now regarded the mean value theorem as only distantly related to his concerns, interrupts his reverie to ask, "What's he riding, man, donkey cart?"

For a moment, Hafez the Intelligent scowls, alarmed, no doubt, by the possibility that Mr. Waldburger's reference to a donkey cart contains an infinitely subtle ethnic slight.

"Now somewhere along the way," I go on gamely, "this character stops to look at an especially vivid desert flower—"

"Guy riding a Harley," Mr. Waldburger asks, marveling at the mystery of it all, "stops to look at a *flower*?"

"Well just suppose," I say. Mr. Waldburger leans back in his chair, which he more or less dwarfs, and arranges his smooth young face in an expression indicating that he has made a large and generous intellectual concession.

"So right after this, he gets back on his bike, opens the throttle and roars off, the open road rushing out like a ribbon, his speed topping out at, *oh,* I don't know. How fast can a Harley go? Mr. Waldburger?"

"Fast as you want it, big machine outaccelerates an F-16 on the runway," says Mr. Waldburger with the kind of easy confidence he never displays in talking about the calculus.

Settling the matter myself, I say, "For that moment he attains a speed of 125 miles an hour."

Hafez the Intelligent looks up at me, understanding at once the whole course of the argument to come.

"So his instantaneous speed at these two moments, first that's 0, I mean when he's stopping to look at the flower, he's not going any-where . . ."

Suddenly alert to the sexist implications of what I am saying, Ms. Ackeroyd asks, "When *he's* stopping?" in a loud, clear voice, prompt-ing Mr. Waldburger to snort derisively from his seat.

"What, you don't think women ride motorcycles?" Ms. Ackeroyd asks in a tone of voice that suggests an almost infinite capacity for bel-ligerence.

"Whatever," says Mr. Waldburger.

" . . . and his or her instantaneous speed at some later moment is 125 miles an hour, and what the mean value theorem tells you, if you think about it—," here I tap the blackboard, with its equation, mean-ingfully, "is that whatever these instantaneous speeds may be, at *some* point in the hour his or her instantaneous speed *must* be exactly 80 miles an hour." I move my hand lamely in the air to simulate a mo-torcycle moving through space.

For a moment, the class contemplates my remarks. Then Mr. Waldburger says with the urgency of discovery, "Got to be, man. Guy's going from 0 to 125, I mean, he's *got* to pass 80 miles an hour. No way he's going to like go from 79 miles an hour one moment, 81 miles an hour next?" He says this with a combative air that suggests that *I* might be arguing the contrary.

Hafez the Intelligent for the moment looks nonplussed. I know that he is searching for a counterargument; and even though *I know* that the mean value theorem is true, I am half-afraid that he is going to come up with an objection I cannot answer.

"Well, yes," I say. "But let me ask you another question. The mo-torcyclist covers eighty miles in precisely one hour . . ."

"This the same guy, right?" asks Mr. Waldburger, concerned about the details.

"Same *person.* Now why couldn't his or her instantaneous speed be 79 miles an hour for the whole time he or she is traveling?"

"Because then their *average* speed would be like 79 miles an hour and not 80 miles an hour," Ms. Ackeroyd blurts out, "like you said."

Hafez the Intelligent scowls. "That is not right," he says.

"How could his average speed be 80 miles an hour if this doofus is going 79 miles an hour the whole time?" Ms. Ackeroyd asks. "I mean, really."

"Why is not possible?"

I interject, saying, "After all, instantaneous speed is defined in terms of the limit of average speeds taken over shorter and shorter intervals of time. Now why couldn't that limit be 79 miles an hour even though the motorcyclist actually covered eighty miles in the hour?"

And from Mr. Waldburger himself comes the entirely unexpected entirely lucid answer. "Can't happen, man."

"Why it can't happen?" Hafez the Intelligent demands, his heavy, spotted brow furrowed.

Raising himself just slightly from his habitual slouch, Mr. Waldburger points to the blackboard. "Mean value theorem," he says. "Mean value theorem tells you it can't happen."

Mutable Meaning

The mean value theorem makes a claim about an entire class of functions, and so a claim about the processes that they represent, and so ultimately a claim about a determinate aspect of the world. It happens often in mathematics that as the corollaries and consequences of a great theorem are spun out, some part of the theorem's meaning mutates as its verbal embodiment changes. So it is with the mean value theorem. The theorem says one particular thing, but it implies many others. It implies among other things that if the derivative of f is 0 on an interval (a, b), then f must be *constant* on (a, b), returning the same value for each argument it addresses. Such is the *constant value* theorem. It is a theorem that makes intuitive sense, of course. The derivative of a function is a sign of that function's change, and if the derivative does not change, the function is going nowhere at all. But between the definition of the derivative and the constant value of a function on an interval, there is only the cold reaches of space: the solid and satisfying conviction prompted by the constant value theorem hinges on the mean value theorem.

Beyond what this little theorem says, there is the radical shift in perspective that it suggests. Up until now, the function has come first in the natural order of things, the derivative of the function appearing in an ancillary role: it is the derivative that has attended the function. This reflects nothing more than the facts of the matter. The function is the essential conceptual object; the derivative is defined *in terms of* the function. And yet the constant value theorem suggests the development of a contrary pattern, a countervailing melody. It is the *derivative* in this theorem that has come to command the function, and this suggests the extraordinary idea that in the derivative mathematicians have discovered an exquisite instrument, one capable by means of the intense light it sheds on a function's *local* character of illuminating in a variety of tones the function's *global* nature.

A significant corollary follows from the constant value theorem, one whose lurid light flickers not only over mathematics but throughout our intellectual experiences. Two functions f and g are introduced—two functions and thus two mathematical objects, but also two processes, two coordinations of space and time, two investments in the real world. The assumption made is that their derivatives coincide on an interval (a, b): $Df = Dg$.[2] It follows that on (a, b) the *difference* $f - g$ between these functions is constant. Whatever the values of f or g,

$$f(t) = g(t) + C,$$

one number serving to mark the difference between these functions throughout an interval. Such is the *differential diversity theorem,* the derivatives again coming to command the functions.

The proof is easy, dramatic evidence that in mathematics what is important is not necessarily what is hard. Thus let F be a new function, one whose purpose in life is simply to measure the difference between f and g.

Start with

$$F(t) = f(t) - g(t).$$

---

[2] The notation Df—read as *the derivative of* f—is often useful as an alternative to Leibnitz's notation df/dt.

Differentiate both sides of this equation:

$$DF(t) = D[f(t) - g(t)].$$

Remember that

$$D[f(t) - g(t)] = Df(t) - Dg(t),$$

so that

$$DF(t) = Df(t) - Dg(t);$$

and remember, too, that on (a, b), $Df(t)$ *equals* $Dg(t)$, so that

$$DF(t) = Df(t) - Dg(t) = 0.$$

But functions whose derivatives are zero on an open interval are constant on that open interval. Since F measures the difference between f and g, it follows that this difference must also be constant:

$$f(t) - g(t) = C;$$

but this is another way of saying

$$f(t) = g(t) + C,$$

which is what the *theorem* says.

A simple function coordinates changes in time with changes in something else. Processes in nature begin at various times, and they begin in various circumstances or places. This is a fact. The world is various. But many processes in nature are more or less alike; amidst the differences in the world, there are similarities, strong patterns, strange resemblances, and this, too, is a fact. Tolstoy chose to concentrate the force of his genius on one small collection of *un*happy families in *Anna*

Karenina, one descriptively manageable set of circumstances. Art is drawn irresistibly to what is singular, but mathematics is drawn irresistibly toward what is *generic.* The novelist is interested in those satisfyingly unhappy characters: Vronsky, his strong teeth aching, the self-absorbed Levin, and Anna herself, progressing inexorably over the course of six hundred pages from being one man's possession to another man's pest. The mathematician is interested in the happy families. They resemble each other, but they are not identical. This suggests a general problem of intellectual accommodation. What intellectual instrument is adequate to the task of describing structures that are similar in a way that is faithful to their differences? Ordinary language achieves this end by means of a certain contrivance, one that finds its perfect expression in moral maxims or proverbs: *As a ring of gold in a swine's snout,* it is written in Proverbs 22, *So is a fair woman that turneth aside from discretion,* a description that in its artful generality could be construed to cover *both* Anna Karenina and Sadie Thompson.

But the calculus treats functions and not families. Things fall toward the center of the earth in Tuscany as well as Trenton, New Jersey, but they fall at different times, and they fall from different heights. No *one* mathematical instrument or function could possibly describe them all with any specificity. On the other hand, no scheme of independent and uncoordinated functions could do justice to the fact that various processes, like various families, are essentially alike. Things fall toward the center of the earth in the same way. Either of two facts thus stands to be intellectually slighted. What is required to do justice to *both* difference and similarity in processes is a set of functions intrinsically alike and yet sufficiently different.

This odd, urgent requirement is met by a *family* of functions differing only by a constant. A family of functions so conceived is adequate to the description of difference.[3] Each particular function is what it is and not some other thing. But it is adequate, too, to the expression of similarity. Each function is different but only to the degree marked by the constant. Faced with the most obvious of facts about nature and

---

[3] Appreciating the force of this insight, the reader—my reader—has just shaken hands with the theory of ordinary differential equations. See chapter 20, pp. 220 *et seq.*

about life, the mathematician invokes precisely *these* functions as his descriptive instruments. The conditions under which such families of functions will be available to the calculus are expressed by the differential diversity theorem; functions agreeing in their derivatives, the theorem states, differ on an interval only by a constant. It is the derivative of a real-valued function that like some pulsing light illuminates again the behavior of the function, enforcing among otherwise anarchic and wayward mathematical objects a stern uniformity of behavior. Such is the proximate burden of the mean value theorem, which is now revealed to play a transcendental role in the scheme of things.

Did Lagrange see to the far corners of the mean value theorem, sensing as his depression leveled and then lifted that he was now among life's happy families, the rich family of human beings in the real world having, as Lagrange himself first demonstrated, an inevitable mimetic analogue among the differentiable functions? For the disturbing, disorienting, slightly mad thought is that *if* a human family can be described by a mathematical function, the happy families *must* be described by a family of functions agreeing over an interval in their derivatives. Of the domestic coziness that Lagrange achieved, the whispered intimacies, the flushed and sudden satisfactions that attend a middle-aged man taking a very young wife, of this, the warm and flowing juice of life, not a trace remains, the cosmic function dipping here to zero and then winking out of existence. And this, too, I suppose, is a part of what Tolstoy meant when he said that happy families are all alike.

The Anatomy Lesson:

Proof of the Mean Value Theorem

Every proof in mathematics returns ultimately to the place where it begins, the difference between the initial conjecture and the settled conclusion, which are, after all, one and the same thing, the intervening matter of decisive *demonstration*. The beginning of the journey and so the initial conjecture is this: that *if* a function f is continuous on the closed interval $[a, b]$ and differentiable on the open interval (a, b), then there exists a point c in (a, b) such that

$$\frac{df(c)}{dt} = \frac{f(b) - f(a)}{b - a}.$$

If the proof begins here it is to this very statement that it must return.

The proof of the mean value theorem is awkward, but the strategic lines of its development are clear enough. If geometrically, the mean value theorem is Rolle's theorem applied to a chord, measure the *vertical* distance between chord and curve, the mathematical tactician suggests, and express that measure as the range of a brand-new function—h, say. Show next that h satisfies the hypothesis of Rolle's theorem and then use the conclusion of Rolle's theorem to enforce the conclusion of the mean value theorem. Awkwardness is the result of indirection. Given a function f, the mathematician must reach over to h, and then return to f.

Whatever the definition of h, it is a function measuring distances between points on the curve $f(t)$ and points on the chord **AB**, specifying a *difference* for each value of t between a, where f commences to be of interest, and b, where f lapses into insignificance.

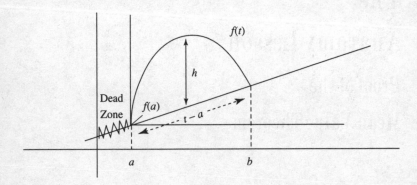

As yet, no way has been introduced to specify points on **AB** itself, and so a preliminary definition of h contains an opaque something at a crucial point:

$$h(t) = f(t) - \text{Something specifying points on } \mathbf{AB}.$$

But a chord is a simple straight line, and straight lines submit to the discipline of an equation expressing their innermost nature: $y = mt + b$, where m is the line's slope and b its y-intercept, the place where the line crosses the vertical axis. In the case of **AB**, that slope is known; it is, by definition, the ratio

$$\frac{f(b) - f(a)}{b - a}.$$

If $f(b) - f(a)/(b - a)$ corresponds to m then $t - a$ must correspond to t itself. The mean value theorem makes an assertion only for the closed interval *between* a and b. Points beyond, the theorem ignores. Substituting $t - a$ for t itself is a way of enforcing that ignorance.

Progress in the project of finding a way to specify the points on **AB** analytically is now two-thirds complete, with the general equation for a straight line $y = mt + b$ receiving partial specification as

$$y = \frac{f(b) - f(a)}{b - a}(t - a) + b.$$

What remains to be fixed is the identity of *b*. Now, as far as the mean value theorem is concerned, **AB** comes into existence at *a* and departs from existence at *b*. This suggests that the distance between *t* = *a* and the *y*-axis is incidental. It plays no role in calculations to come. The space that it spans is dead. It may be canceled by simply moving the *y*-axis so that it comes to rest at *t* = *a*.

That having been done, it follows that it is *f*(*a*) itself that expresses the *y*-intercept of the chord, and the project of invigorating

$$y = mt + b$$

is completed by the formula

$$y = \frac{f(b) - f(a)}{b - a}(t - a) + f(a).$$

The requisite definition of *h* follows:

$$h(t) = f(t) - \left[\frac{f(b) - f(a)}{b - a}(t - a) + f(a) \right].$$

When the minus sign is allowed to exert its baleful influence on the addition sign within the brackets, this formula has a useful variant:

$$h(t) = f(t) - f(a) - \frac{f(b) - f(a)}{b - a}(t - a).$$

Two facts about *h* must now be invoked. First, *h* is continuous on [*a*, *b*], and second, *h* is differentiable on (*a*, *b*). These facts must be demonstrated in their turn, but their demonstration is trivial. The function *h* is continuous on [*a*, *b*] because it is the sum of *f* and a polynomial, and it is differentiable on (*a*, *b*) for the same reason; if this seems to lack the force of direct demonstration, then let those facts enter the discussion simply as facts, with their status in the proof resting on my say-so.

Given the way in which *h* was constructed, it is obvious that at the points *a* and *b*, *h* must be 0, since there is *no* difference between the curve *f*(*t*) and the chord **AB** at the very places that they meet. This is evident from the picture; it is evident again from the definition of *h*. Thus at *a*,

$$h(a) = f(a) - f(a) - \frac{f(b) - f(a)}{b - a}(a - a) = 0;$$

and at b,

$$h(b) = f(b) - f(a) - \frac{f(b) - f(a)}{b - a}(b - a),$$

which, when $b - a$ is canceled from the equation, reduces to

$$h(b) = f(b) - f(a) - f(b) - f(a) = 0.$$

It follows that $h(a)$ *equals* $h(b)$, the two assessments of h at a and b collapsing ignominiously into nothingness.

The new function h is continuous on $[a, b]$ and differentiable on (a, b), and what is more, $h(a) = h(b) = 0$; h thus satisfies the hypotheses of Rolle's theorem. From Rolle's theorem, it follows that there must be a number c somewhere in (a, b) such that

$$\frac{dh(c)}{dt} = 0.$$

The gate of conviction now begins to swing. An equation is an exercise in symmetry and balance; two things exist in a precarious state of equipoise whenever an equation is affirmed. The equation

$$h(t) = \left[f(t) - f(a) - \frac{f(b) - f(a)}{b - a}(t - a) \right]$$

implies that derivatives of both sides of the equation must be equal:

$$Dh(t) = D\left[f(t) - f(a) - \frac{f(b) - f(a)}{b - a}(t - a) \right].$$

To evaluate $Dh(t)$, it is necessary only to evaluate

$$D\left[f(t) - f(a) - \frac{f(b) - f(a)}{b - a}(t - a) \right].$$

Considering the complexity of the expression within brackets, this may seem a daunting task. But one of the secondary miracles of the calculus is that it provides for the first time in intellectual history a set of procedures, akin really to algorithms, which if properly understood expedite the tedious and difficult work that would otherwise figure in the determination of this derivative. The expression to be differentiated consists of two parts (which I have enclosed in brackets):

$$[f(t) - f(a)] - \left[\frac{f(b) - f(a)}{b - a} (t - a) \right].$$

Each part has a simple structure in turn, the first expressed as a subtraction:

$$[f(t) - f(a)] = f(t) - f(a),$$

and the second as a product:

$$\left[\frac{f(b) - f(a)}{b - a} (t - a) \right] = \frac{f(b) - f(a)}{b - a} \times (t - a).$$

What follows is mechanical. The term $f(a)$ is constant in its values. The derivative of a constant function is 0; $f(a)$ vanishes from further consideration.

With nothing known about $f(t)$, *its* derivative is recorded as $df(t)/dt$, a way of signifying its derivative *whatever* it is.

The ratio $f(b) - f(a)/(b - a)$ is constant, and so the product of $f(b) - f(a)/(b - a)$ and $t - a$ follows the rule of differentiation that affirms that the derivative of a function of the form $f(t) = at$ is simply a, yielding

$$\frac{df(t)}{dt} = \frac{f(b) - f(a)}{b - a}.$$

The seemingly formidable task of differentiating both sides of the equation

$$h(t) = \left[f(t) - f(a) - \frac{f(b) - f(a)}{b - a} (t - a) \right]$$

yields

$$\frac{dh(t)}{dt} = \frac{df(t)}{dt} - \frac{f(b) - f(a)}{b - a}.$$

This is how things are in general, with t acting as a variable. At the particular point c, this equation reads

$$\frac{dh(c)}{dt} = \frac{df(c)}{dt} - \frac{f(b) - f(a)}{b - a}.$$

Rolle's theorem establishes—it has *already* established—that the derivative of h at c is 0, so that

$$0 = \frac{df(c)}{dt} - \frac{f(b) - f(a)}{b - a}.$$

The familiar ratio

$$\frac{f(b) - f(a)}{b - a}$$

is now added to both sides of this equation, gracefully yielding

$$\frac{df(c)}{dt} = \frac{f(b) - f(a)}{b - a},$$

but *this* is the desired conclusion of the proof, the affirmation of the mean value theorem, and the place whence the journey began.

No step in this proof is difficult, but overall the impression of complexity is considerable.

chapter 20

The
Song of
Igor

I LIKED THEM, MY STUDENTS. ONE MUSCULAR YOUNG MAN, WITH untroubled eyes and a clear complexion, was painfully nervous about speaking aloud, his smooth face mottling with blood whenever I called upon him. By and by, he relaxed. He was a member of the California Highway Patrol. He told us stories of the road, all delivered in that rich police vocabulary, full of *perps* and *peds* and accounts of how one day chasing a perp at speeds in excess of one hundred miles an hour easy this ped he just walked in front of the vehicle and it wasn't even a *bam,* you know, more like a *splat* and yeah, you get used to it. When I asked him why he had not volunteered for the motorcycle patrol, given his love of speed and danger, he replied solemnly that my mom she won't let me, she'd just freak out. There were California lower-class natives in my class, monstrous boys with ripe pustules and bleached blond hair, an earring in one ear, or young women with thick legs and a kind of low-slung pelvis, or the statistically inevitable one-in-a-thousand raven-haired beauty, eyes glowing, lips cherry red, as tart as tart could be. They lived at home, these teenagers, with their parents, rootless themselves, a grandmother yet living in Idaho, in a house by a field where the wind blows off the plains, the screen door banging fitfully, someone calling *Rachel* over and over again. They ate Cheez Whiz, Big Macs, Doritos, mayonnaise on white bread, Velveeta, macaroni and cheese, fudge pies; they aspired to police work or accounting or business management or *I dunno, get a job, something, I guess;* they worked part-time, and took care of their baby brothers, their mothers not entirely diligent about birth control, leaving the diaphragm in the medicine chest and prone thus to pregnancies in their forties; they did not read and could not write and they were touching and earnest and good-natured, curiously old-fashioned, devoted to those mothers of theirs whom they referred to as Mom—*I think my mom she's so neat.*

The department of mathematics, on the other hand, was rather a malignant place, the atmosphere of intimidation characteristic of such institutions conveyed by posted homework assignments designed to stupefy even capable students and an enormous wall chart depicting famous mathematicians in poses suggesting (in pictures like those on post-office bulletin boards) ineliminable weirdness in each and every one, the great Gauss emerging wall-eyed from the chart and Newton himself, his wig askew, looking as if he had risen with a smashing headache. Several doors into the office suite, the chairman of the department reposed like a spider. He was completely bald, his head mounted on his narrow shoulders like an egg, utterly officious, and intolerably burdened by his position, or so he told me on the occasion of our single interview, sweeping the room with his small hand as if to say in a tone of infuriated resignation *look at all this,* the paperwork involved in managing the department piled on his desk and even overflowing to the floor below.

The department was additionally home to a number of eccentric eccentrics. One lunatic spent his time writing science fiction novels in which strangely voluptuous women found themselves chained to the thirteenth dimension in their underwear. Another raced motorcycles professionally and was fond of appearing in class in a tight nylon body suit and helmet. A few Russian immigrants had come to the university by means of fantastic journeys, inconceivably difficult stratagems. There was Igor M, for example, the M trailing off into a series of sibilant but unaccommodating Cyrillic sounds, so that rather than risk his indignation at my florid mispronunciation of his name, I took to greeting him in the second person Californian—*Hey Igor, how you doin'?*, the sheer absurdity of being addressed as Igor for the first time in his adult life sufficient to cancel out the implied impudence. He never seemed to mind and after a time took to responding in fashion: *Iss doing fine*.

Igor was no more than five feet tall; he had an unhealthy look, evoked as much by a life of fear as by a bad diet; he spoke English with an ineradicable, almost opaque Russian accent, and he regarded his colleagues and students, now that he had miraculously been awarded tenure, with barely disguised and entirely condign contempt. He was a powerful and learned mathematician, trained in the Moscow school, a student of a student of the great Kolmogorov. His courses were of fearful difficulty. "Iss nofink," he would say when it was observed that he had again failed almost all of the students reckless enough to study the calculus with him. The chairman of the department endeavored to interpose himself between the implacable Igor and his enraged students. "Now look, Igor," he would whine placatingly, "you have to remember that this isn't MIT."

"Iss not possible forget," Igor would intone.

And there matters would stand, so that in the end Igor had no students whatsoever and remained free to devote himself entirely to his research in partial differential equations. He had lost eight years altogether. When he had applied for permission to emigrate, Russian authorities had stripped him of his position at the University of Moscow; he had sat at home in his tiny apartment. "No books," he said, "no paper. Nofink." Four years had been swallowed up learning his own brand of barely serviceable English. Now that he had put teaching behind him, he worked furiously to catch up. "Vary difficult," he said. "So much *ch*eppens."

Differential Traces

In life and in nature it sometimes *ch*eppens that things may be reversed. Stepping forward, I step back, the two steps, one forward and one back, canceling one another so that after they have been completed, I am where I started, having done something but accomplished little, a useful metaphor for a great many activities in life. The shuffling pattern of versal and reversal is on hand in mathematics. The elementary arithmetic operations, for example, do and undo one another. Subtraction undoes addition, so that $2 + 7$, a step forward, is undone by $(2 + 7) - 7$, a step backward. But then again, division undoes multiplication, so that 12×3 when divided by 3 is just 12, the place where the two operations began.

These familiar (but not trivial) facts make possible the legerdemain of elementary algebra, as when

$$\frac{x^2 - 1}{x - 1}$$

is factored into

$$\frac{(x + 1)(x - 1)}{x - 1},$$

the revision indicating immediately that the original expression was composed by first multiplying $x + 1$ by $x - 1$ and then dividing the result by $x - 1$. It is the fact that division and multiplication are reversals of one another that allows the mathematician to strike $x - 1$ from the numerator and denominator of the expression, leaving only $x + 1$ behind, so that $(x^2 - 1)/(x - 1)$ and $x + 1$ are, if $x \neq 1$, suddenly seen to be the *same*, an example that illustrates that the mysterious power of algebraic techniques resides on occasion in something so simple as the left hand undoing what the right hand does first.

Up until now, differentiation has been entirely a forward-looking operation. *Differentiation,* as in the activity of taking limits, the particular number 4 serving as the derivative of the function $f(t) = t^2$ at the point $t = 2$; but differentiation also, as in the activity of pairing one function t^2 to another function, $2t$, and so pairing one process with another, the *activity* achieving a connection between two ways of coordinating time and space. Intellectual motion proceeds in one direction. A function is given and assessed at a point: one passes to the number representing its derivative. Or a function is given generally, with no regard for any particular one of its arguments: one passes to the function representing its derivatives. The position function $P(t)$ indicates distance. *Look ahead!* There is the velocity function **vel**(t), the derivative of $P(t)$, coming up like thunder out of China in the east. And yet the pattern of reversal that is so conspicuous a feature of elementary arithmetic and algebra is on hand as well in the calculus. It is, in fact, the characteristic mental motion of the calculus, the subject's great gesture.

Everyone knows what an equation is. An equation is an expression affirming that two or more things are equal, ordinary English expressing ordinary equations by means of an ordinary copula—*Benjamin Franklin is the man who invented bifocals*. Equations in mathematics acquire their great usefulness when they contain an *unknown* something, the equation then serving as a means for the identification of this unknown; but before the air of remembered dismay at word problems met and missed becomes too rank, let me offer this conciliatory thought: that so simple a sentence as "He is tall, dark, handsome, wears a thin mustache, and played opposite Vivien Leigh in the film *Gone With the Wind*" also succeeds in the specification of something unknown by means of an adroit compilation of associated verbal clues and conditions. What is unknown in this sentence is who *he* is, and the sentence serves to identify the unknown by setting out certain conditions that *he,* the unknown one, must meet. In this case, the conditions are obvious enough to identify the character in question: *he* is Clark Gable.

The provocative (and ubiquitous) pattern in which an unknown something is balanced against certain presumptively identifying conditions reappears in elementary mathematics. The unknowns are very often numbers, as in the equation $5x = 25$, which says simply enough that an unknown number when multiplied by 5 happens to equal 25. It is the extraordinary achievement of elementary algebra that it affords the mathematician the opportunity to *manipulate* the equation so as to identify in a flash the value of the variable. Divide both sides of the equation by 5, the algebraic rules say, and *lo,* there is the answer, clear as sunshine: *x* is 5. It is the fact that $5x = 25$ is an *equation,* and so a specification that two things are the same, that makes the double division legitimate; it is the generality inherent in the use of variables such as *x,* which ordinary language captures in its pronouns, that makes the double division successful.

The examples offered by elementary algebra are often uninspiring if only because no one wishes really to know which numbers correspond to the unknowns, the unknowns in word problems referring always to a strangely meditative farmer standing forlornly on that illustrated textbook hill of his, wondering in a way that suggests nothing of the power of mathematics how many turnips he might grow if he had two tons of fertilizer. Recalling my own experiences, let me blurt out that God knows I *hated* teaching elementary algebra, especially to the roomful of adults fully prepared to meet word problems with their own well-developed, coruscating sense of the absurdity of the examples. But as always in any great art, the matter and the method need not necessarily be the same at first, the matter trivial (farmers and their fields), but the method, in the case of elementary algebra, suggesting in incremental steps the power of a system of equations adequate to nothing less than the description of the world.

The concept of a function gives to mathematicians an instrument of supreme physical relevance, one obviously intended to capture the coordination of time and space, and one thus intended by the world's architect to represent things in nature. A complete catalog of physical processes would reveal the way in which the world works and so present the mathematician an intimate view of the mind of God. No such catalog, I regret to say, is currently available, and none is in prospect. Nature presents itself to the mathematician (and everyone else) as a problem, a series of vexing riddles. But in the context of the calculus (and so the context of science itself), problems and riddles alike take on a certain distinctive shape, a suggestive profile. Far ahead, there is an unknown function, one that *might* bring about a specific correlation of time and space, a discernible connection between position, say, and time, or a discernible connection between any two spatial and temporal aspects of experience. When I say "specific" I mean only that its identity having been uncovered, the function stands revealed as a particular mathematical creature, the mathematician able to say that *it* is exponential or trigonometric or logarithmic or otherwise complicated, but still recognizable as a part of the collection of *his* familiar descriptive instruments. Although the function is initially unknown, in an equation it yet answers to a certain description, just as *x* answers to a certain description as the number which when multiplied by 5 is 25. Like sunken and encrusted Spanish treasure, the function is something that by means of contemplative effort might be withdrawn from its inky epistemological depths, the withdrawal accomplished, if at all, by an artful combination of mathematical methods designed in combination to force the function's identity. Even those bright but baffled housewives taking mathematics in order to move from Milpitas to medical school—I may have inadvertently multiplied one such housewife in my experience in order to populate an entire roomful of discontented and ambitious women—should sense, with whatever reluctance, the extraordinary act of intellectual audacity embodied in the equational art, its sheer contrivance, a function, and so an abstract representation of a process, hinted at and finally hunted down by means of a collection of verbal constraints, a form of words.

The Sound of My Own Voice

Whenever they are called upon to evaluate me in secret, my students never fail to observe that *Mr. Berlinski, he really likes to hear himself talk.*

I am talking now, enjoying the sound of my own voice. I have been going on and on. Igor M, who has been asked by the chair to observe my classes, drums his fingertips impatiently on the wooden surface of his desk, the only person in the room actually to be dwarfed by the children's seats.

What descriptions, I ask, *are* relevant to the specification of an unknown function? There is no one answer to this question, but in the calculus the most relevant of descriptions mentions its derivatives and so identifies the unknown function by its differential traces. *Yes, this is the idea—the world contains differential clues, a series of signs*. It is in this way that differential equations arise, their discovery in the seventeenth and eighteenth centuries offering mathematicians the prospect of having suddenly entered a dazzling new world of subtlety and imaginative power, and their discovery comes again to life whenever the printed page is laid flat and the bewitched reader blows the dust from the centuries and allows those symbols to acquire their dense meaning.

I pause to survey the effect this somewhat florid speech is having on my class.

A function F is given or carelessly invoked, its identity shrouded, opaque, black as the inky night. All that is known of the function is that it *is* a function, an instrument of coordination. That and the fact that when differentiated, *this* function yields the function $3t^2$. The clue, and the implicit question that prompted it, may be arranged as a differential equation:

$$\frac{dF}{dt} = 3t^2.$$

The equation repeats the verbal flourish that introduced it. Some unknown function F, the symbols say, when differentiated yields the function $3t^2$. What is wanted is the identity of F; the equation provides the clue. Nothing more. In its overall form, this equation is kin to algebraic equations such as $5x = 25$, the difference only that here x answers to a number and up there in the differential cockpit, F answers to a function; but differential equations, even on this level, have an air of cold command that is missing entirely from algebra.

In the case of this equation, intuition and the remembered rules of differentiation suffice to pin down the identity of F with satisfying ease. The function F is, it *must* be, the function $F(t) = t^3$. The proof? Just that when differentiated, t^3 is $3t^2$:

$$\frac{dt^3}{dt} = 3t^2.$$

The derivative of *any* function of the form $F(t) = t^a$ is always at^{a-1}.

Beyond the specifics of this example, an old mental muscle may be observed, twitching and at work. The identification of F proceeds by means of its description as the function whose derivative is $3t^2$. Proceeding from F to its derivative, the mathematician is proceeding *forward*. But the solution of the equation turns on a reversal in direction. Proceeding from $3t^2$ to an identification of F, the mathematician is proceeding *backward*, toward the original function. If the forward motion embodies differentiation, the backward motion embodies *antidifferentiation*, the function $F(t) = t^3$ serving as an antiderivative of the function $f(t) = 3t^2$. It is antidifferentiation that endows the formerly opaque F with its identity as a very particular function.

All at once, Igor M interrupts the pleasant sense I have of talking to myself.

"You give now formal definition," he says.

"Very well. A function F," I write on the blackboard, "is an anti-derivative of f . . ."

A certain current of confusion persuades me that I have for the moment lost everyone but Igor M, who now sits beaming in his seat.

"What's the matter?" I point to F with the tip of the chalk and to f. Hafez the Intelligent puts his finger on what is troubling.

"These are two different functions?" he demands.

"Well, yes," I say. "Two *different* functions."

"Then why you call both F?" and in a flash I see how students unused to symbolism might find the distinction between lower-case f and capital F problematic, the professional choosing a slight variation in symbols just to hint that the two functions are related, students regarding the relationship as unformed and so the symbolism as confusing.

I look to Igor M for a moment; it is plain that he regards Hafez the Intelligent as an imbecile; I am not sure that he does not regard me in the same way. Later he tells me that it is a pedagogical mistake to take questions from my students. "Ask question of Kolmogorov?" He circles his finger around his temple to signify incipient madness.

"At any rate," I say gamely, caught in the glare of two stares, "this function F"—and here I tap the blackboard—"is an antiderivative of f if for every t, the derivative of F at t just is $f(t)$."

Igor M says, "Iss not right," loudly. My loyal students turn to look at him, astonished to find me in the position they generally find themselves.

"Why not right?" Hafez the Intelligent asks belligerently.

"Now lewk," Igor M says patiently, "iss not right becawse he fawged to say for every *t* in *domain* of *f*."

And, of course, I had.

"Fussy, fussy," says Ms. Klubsmond, in a low but carrying voice; she has always regarded as mean-spirited my attempts to persuade her of the virtues of simple but carefully drawn definitions, preferring always in class to refer to the class of continuous functions as *you know, those things*. Now the hour of her triumph is at hand.

"No, no, Professor M is right," I say generously, although I wish to drive a stake through his knowledgeable Russian heart. "*F* is an antiderivative of *f* if for every *t in the domain* of *f*, the derivative of *F* at *t* is just *f*(*t*)."

Letting these niceties lapse, I remark that antidifferentiation has a simple symbolic meaning. If *F* is an antiderivative of *f*, then—and here I write the formulas on the blackboard with the flat side of the chalk so that they seemed shadowed:

$$\boldsymbol{D}F(t) = f(t)$$

or

$$\frac{dF(t)}{dt} = f(t).$$

This is something true by fiat. Meaning has been given to symbols, and in this case the symbols have nothing to say about it.

It remains for the mathematician to supply a notation for the process of antidifferentiation, and this the calculus provides. Suppose that *f*(*t*) is a continuous function. The elegant and shapely symbol

$$\int f$$

serves to denote its antiderivative.

The bell signaling the end of the hour is about to sound. "Anti-differentiation is an operation that involves a reversal of form," I say, "and if a pictorial image is wanted it should be drawn from the world of fencing, as when the fencing master thrusts—*differentiation*—and with an enigmatic smile playing on his features after his opponent murmurs *touché,* backs up and retracts his elegant foil—*antidifferentiation.*"

From the corner of my eye, I can see Igor M making a delicate, diffident thrusting motion with the loosely closed fist of his right hand, his thumb resting gently on his forefinger.

And Yet Again the Mean Value Theorem

Differentiation goes from a function to a function; ditto, anti-differentiation. But with a difference. Differentiation goes from a function to a specific function, an overwhelmingly particular Other, $2t$ in the case of $f(t) = t^2$. Differentiation is mathematically a puritanical operation. Not so antidifferentiation, which retains something of the promiscuous. If $F(t) = t^3$ is an antiderivative of $f(t) = 3t^2$, then so is $F(t) = t^3 - 5$ or $F(t) = t^3 + 217$. This peculiar promiscuity follows from the most elementary facts of differentiation. The derivative of a sum, the rules of differentiation reveal, is the sum of its derivatives, and the derivative of a constant is 0. This suggests, the inference almost immediate, that the antiderivative of $f(t) = 3t^2$ is actually a *family* of functions $F(t) = t^3 + C$, with C taking on varying values—with C just about anything you please.

On the other hand, *no* functions outside this family are suitable antiderivatives for $f(t) = 3t^2$, a fact that follows directly from the mean value theorem, which in one of its many prior incarnations revealed that if two functions have the same derivative, as in the case of $F(t) = t^3 - 5$ and $F(t) = t^3 + 217$, then they must differ only by a constant. Differentiation thus goes from a function to a function, in the most general of circumstances, but antidifferentiation goes from a function to a family and *only* to a family of functions. The crucial facts now admit symbolic expression. *Antidifferentiation reverses differentiation:*

$$\int \frac{dF}{dt} = F(t) + C,$$

the symbols saying that if the function F is differentiated, and then antidifferentiated, the result will be the same function taken in a family way.

But equally, *differentiation reverses antidifferentiation:*

$$\frac{d \int f}{dt} = f,$$

the symbols now saying, go ahead and find the antiderivative of some function f, then differentiate *that;* the result will be f again—the function f and not the family, please note.

These formulas are *less* perspicuous than their purely vernacular explanation: they do not breathe. They are instead fabulously compact ways of presenting information, and in time their eerie concision comes to appear as a form of beauty.

The class tenses at my claim that in time these symbols will seem beautiful.

"You want to give yourself plenty of time there," Mr. Waldburger genially affirms.

The elementary functions whose derivatives have already been specified yield a set of functions whose antiderivatives are *necessarily* known. If the derivative of t^2 is $2t$, then the antiderivative of $2t$ must be $t^2 + C$. This makes for a chart of antiderivatives, one entirely coordinate to the original chart specifying the derivatives of the elementary functions.

Derivatives	Antiderivatives
$\frac{d}{dt}[C] = 0$	$\int 0 = C$
$\frac{d}{dt}[at] = a$	$\int a = at + C$
$\frac{d}{dt}[af(t)] = af'(t)$	$\int af(t) = a\int f(t)\ dt$
$\frac{d}{dt}[f(t) \pm g(t)] = f'(t) \pm g'(t)$	$\int [f(t) \pm g(t)] = \int f(t) \pm \int g(t)$
$\frac{d}{dt}[t^n] = nt^{n-1}$	$\int t^n = \frac{t^{n+1}}{n+1} + C,\ n \neq -1$
$\frac{d[e^t]}{dt} = e^t$	$\int e^t = e^t + C$
$\frac{d[lnt]}{dt} = \frac{1}{t}$	$\int \frac{1}{t} = ln|t| + C$
$\frac{d}{dt}[\sin t] = \cos t$	$\int \cos t = \sin t + C$
$\frac{d}{dt}[\cos t] = -\sin t$	$\int \sin t = -\cos t + C$

The reciprocal relationship between derivative and antiderivative is a matter of definition, and definitions, according to common cliché, are arbitrary. The very word *definition* compels the belief that no matter how complicated or delicate a definition may be, as in the case of the definition of a limit, it does not, it cannot, grasp the real world and must remain forever a part of the suspiciously closed circle of words chasing words. And yet the definitions of the calculus succeed brilliantly in breaking free of that weary wheel of words. The concept of antidifferentiation reverses the effects of differentiation; it provides as well a sensuous notation in the shapely symbol for antidifferentiation. The idea is a powerful one, and a sensuous notation adds to the stock of things in the world that are lovely to look at. But this suggests a limited landscape: what I am pointing to (and taking on faith) is a panorama.

The Many and the One

The differential equation

$$\frac{dx}{dt} = 3t^2 - 1$$

says that some unknown function, now denoted x, has $3t^2 - 1$ as its derivative. *Find x*, it demands. It is an injunction no longer at the dark margins of incoherence; finding x is a matter of antidifferentiation, with

$$\int 3t^2 - 1$$

denoting the entire family of functions that satisfy the peremptory request. In fact, $\int 3t^2 - 1$ is equal to $F(t) = t^3 - t + C$, as differentiation reveals, the revelation coinciding with the moment in human history when a mathematical system acquires a stunning new power, the ability by means of differential equations to treat functions, with all their compelling power to represent time and space in a real world, as if they were algebraic unknowns, like the x standing in for the farmer's turnip yield. The enlargement in scope is justified by a scheme in which the unknown functions are actually tracked down and identified. It is here, at this very moment when the first utterly trivial differential equation is solved, that the secret form of words is revealed that makes modern science possible.

Like a difficult work of the graphic arts in which what seems simple conceals a world of vibrant depth, the examples to which I am attending yield their riches slowly. Corresponding to the family of functions, $F + C$ is a family of curves in a Cartesian coordinate system, each a vertical translation of the other and each a living record of time passing and space changing.

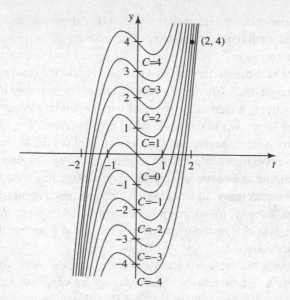

These curves are vertical translations of one another because any two, *f* and *g*, say, differ by a constant *C*, and the constant, being a fixed number, acts simply to raise or lower *f* until it coincides with *g* (or vice versa). Like the functions that they express, the curves constitute a *family,* with the curves sharing a basic similarity in shape and yet distinguished from one another by their varying relationship to the coordinate system. But with the graphical invocation of this family, we reach backward to an old affirmation, for it was the underlying promise of the mean value theorem that mathematics would make available instruments sufficiently flexible to be in their overall respects alike and in their particulars quite different.

Here on the page is the payoff to that promise, the metaphysics emerging from the mathematics, and yielding a wonderful sense of what precisely it might mean to speak of unity in diversity and diversity in unity. The family of curves *as a whole* constitutes the general solution to the differential equation

$$\frac{dx}{dt} = 3t^2 - 1,$$

but each curve is what it is and not some other thing and so a specific record and a particular expression of how time passes *now* and how space changes *here*.

There is in this picture, then, a profound understanding of the nature of change itself. Change is embodied in and represented by a family of functions, a clan of curves. But why *profound*? What lends to this vision of the world the depth that I am claiming? The history of philosophy is littered with countless striking but sterile images. There is Plato's cave, for example, with its dimly dancing shadows; but for all its haunting and provocative beauty, *this* image has gone nowhere in two thousand years, a fact of which I am reminded regularly, when I teach philosophy, by an insidiously nagging inner voice saying *so what?* in a way that instantly provokes a real voice from a real student to an affirmative echo.

But with the mathematician's picture, and its underlying commitment to a certain view of change, things are very different, an insolent *so what?* silenced by the fact that this picture is infinitely productive. Beyond what it shows, it suggests a way of going on—the distinctively *scientific* way of going on.

Falling objects fall toward the center of the earth. This familiar law of nature satisfies a legitimate urge toward the largest description of things. But processes in nature are often interesting precisely to the extent that they are overwhelmingly particular. The cannon is shot from that hill at the *beginning* of the battle, not *later*, and from the *hill* and not the *valley;* the moon appears behind the clouds in the evening and *not* the morning; and the masked figure of death comes to fetch the ashen prince and his great crowd of revelers and wastrels in Poe's story *now* and not *then* and *here* and not *there*.

Given the picture showing a family of curves and thus a congregation of processes, is there a way to specify a particular process, to invigorate the local and the individual? It is a part of the productivity of the picture that this is a question the picture itself makes possible, if only because the picture and the mathematician's analytic apparatus provide for the first time the distinctions required to ask and then answer the question.

The differential equation has by itself an easy nature: it is satisfied by *any* of its solutions and so by any function of the form $F + C$. And this, as the mean value theorem reminds us, is as it must be. A particular solution to a differential equation comes about when the general equation is tied to specified and local circumstances of time and of space.

The mathematician stabs with a stiffened index finger at the interior of a Cartesian coordinate system. What happens if things start *here* and *now?*—the here and now of the question corresponding to the place his moving finger loiters. Say that here and now corresponds to the point <2, 4>. These are the point's coordinates, 2 fixing the time, and 4 the distance; these particular coordinates function as the *initial conditions* imposed upon the differential equation

$$\frac{dx}{dt} = 3t^2 - 1.$$

Initial conditions? Igor M has left, leaving the class to grumble at will. Meaning what? Lower that hand, Ms. Klubsmond, *please;* I *will* open the dark shade to let a single shaft of sunlight shine through; but let me say this, *please*, that things are *easier* than they look and that only the thickness of the finest India paper separates you—yes, *you*— from understanding everything.

The differential equation? That has a general solution in the form of a specified function:

$$F(t) = t^3 - t + C.$$

This we *know*, the knowing already achieved by having trustingly followed the rules for differentiation.

Very well. It is $F(t)$ that we want, $F(t)$ that will tell us where things are going and where they have been; it is $F(t)$, after all, that is a function and so an instrument for the depiction of change. The equation $F(t) = t^3 - t + C$ reveals the general form that $F(t)$ takes, but it fails to distinguish among solutions, treating all of them alike, the very mark of this indiscriminateness that ubiquitous constant C.

And yet we know as well—*isn't this true?*—that whatever it is that is of interest *starts* at the point <2, 4>. We know, that is, that $F(2) = 4$, F acting here as it acts everywhere to send its arguments to its values. But since

$$F(t) = t^3 - t + C$$

in general, it follows that as t takes on the value 2

$$F(2) = 2^3 - 2 + C = 4;$$

and from the fact that $2^3 - 2 + C$ equals 4 only if C is -2, it follows again that in these circumstances C must be -2.

But look what we have done. The general solution to the original differential equation $F(t) = t^3 - t + C$, together with a specification of particular initial conditions $F(2) = 4$, has yielded the particular solution

$$F(t) = t^3 - t - 2.$$

It is *this* F—our F, now, Ms. Klubsmond—that describes, that *expresses*, one curve among many and so specifies one possible process among a family of processes, revealing as it does an absolutely extraordinary method of inquiry, one which like no other method is able to combine in the single efflorescent instrument of a differential equation a device in which the particular is seen as an aspect of the general and the general seen as an expression of the particular.

Ms. Klubsmond takes this all in. "Cool," she says.

Mathematics Yields a Law of Nature

Acceleration is a roller coaster or the blue waters of a swimming pool surging up to meet the diver recently toppled from the high board; but acceleration is *also* one of the concepts of the calculus and so a concept with a fixed mathematical meaning, a role in the scheme of things. The function $f(t) = t^3$, the rules of differentiation tell us, has a derivative in the function $3t^2$, one function giving way to another. But the function $3t^2$—is there any reason that *it* cannot be differentiated as well? If the derivative of $f(t) = t^3$ is $3t^2$, then the derivative of $3t^2$ is simply $6t$, a brand-new mathematical object and indisputably a new mathematical function. Such is the *second derivative* of $f(t) = t^3$.

The second derivative of a real-valued function has a dramatic physical interpretation. The function marking the position of an object tracks distance against time. Its derivative yields instantaneous speed. The second derivative of the very same function must therefore yield its instantaneous change in speed. And this is nothing other than the familiar concept of acceleration, the rate at which speed (and not distance) is changing. If speed takes units expressed, say, as miles per hour, acceleration takes units expressed as *(miles per hour) per hour*— the ratio of speed to time. In all this, the mathematics is faithful to the pattern of development already in place, the concept of acceleration simply an extension of the definition of the derivative and not something entirely new, speed and acceleration appearing in the calculus as concepts of experience that have undergone the peculiar transmogrification of ordinary concepts that find themselves invested with mathematical meaning.

And yet it is in the play between a single physical fact and the concepts of the calculus that the first of the great laws of physics is forged. The *acceleration* of an object—*any* object, from baseballs to ballerinas—moving toward the earth in free fall is a constant g—roughly -32 feet per second in most places. There is a crucial distinction of language here. The physicist speaks of acceleration as a constant; the mathematician, of acceleration as a constant *function*. If acceleration is expressed by a function $\textbf{acc}(t) = g$, this means that one of the antiderivatives of $\textbf{acc}(t) = g$ must be the function $\textbf{vel}(t)$ expressing velocity. Acceleration is *defined* as the derivative of speed.

The general antiderivative

$$\int \textbf{acc}(t)$$

of the constant function $\textbf{acc}(t)$ is a family of velocity functions, but the identity of $\textbf{vel}(t)$ remains opaque. It betokens speed, to be sure, and so generates acceleration, but what is its mathematical form? Antidifferentiation suffices to determine the answer. The hidden identity of **vel**(*t*) is $gt + C_1$.[1] The proof? Just that the derivative of $gt + C_1$ is g itself and **acc**(*t*) = *g*. But what a remarkable conclusion this is. The *mathematical* operation of antidifferentiation has revealed that the speed of a falling object is determined by a simple function **vel**(*t*) = $gt + C_1$, the calculus reaching out now to determine in part how the world is run.

The operation just conducted may be conducted again, almost in the same language. The definition of velocity proceeded in pages past by taking the derivative of a function marking change in position as a function of time. One of the antiderivatives of velocity must be the position function *P*(*t*).

The general antiderivative

$$\int \mathbf{vel}(t)$$

of the velocity function is a family of position functions, but the identity of *P*(*t*) remains opaque. It betokens position, to be sure, and so generates velocity, but what is *its* mathematical form? Antidifferentiation suffices to determine the answer. The velocity function **vel**(*t*) has already been given an identity as $gt + C_1$. The antiderivative

$$\int gt + C_1$$

of a function of the form $gt + C_1$ must in turn have the form[2]

$$\frac{1}{2}gt^2 + C_1 t + C_2,$$

since *this* function, when differentiated, yields $gt + C_1$.[3] This is again a remarkable conclusion. The *mathematical* operation of antidifferentiation has revealed that the speed of a falling object is determined by a simple function **vel**(*t*) = $gt + C_1$, and revealed again that the position of a falling object is determined by another simple function

---

[1] The constant is indexed as C_1 because another constant is coming.

[2] The second constant, C_2, has come.

[3] The derivative of C_2 drops away as 0. The derivative of $C_1 t$ is just C_1 itself. And the derivative of $(1/2)gt^2$ reduces simply to gt. (Remember that the derivative of x^2 is $2x$.) So the derivative of $(1/2)\, gt^2 + C_1 t + C_2$ is $gt + C_1$, as advertised.

$$P(t) = \frac{1}{2}gt^2 + C_1t + C_2,$$

so that not only an object's speed but its very place in the world has been determined by one and the same abstract operation.

And the two constants C_1 and C_2? They arise as the detritus of the mathematical operations. Acceleration is the second derivative of position. In order to capture position from acceleration, antidifferentiation must be performed twice. Each performance yields a constant. But the constants have a *physical* as well as a mathematical meaning and so provide another instance in which mathematics reaches out vigorously to make contact with the real world. The black arts just celebrated reveal that **vel**(t) takes its earthly incarnation as $gt + C_1$. If no time whatsoever has elapsed, **vel**(0) is just C_1 itself, gt vanishing in the annihilation of multiplication by zero. This observation suffices to endow C_1 with its physical identity as the *initial velocity* of a falling object. A diver approaches the end of the high board, gathers herself into a crouch, springs upward, and at the apogee of her splendid spring, tucks her body into a ball, after which she falls downward toward the pool and toward the center of the earth. It is C_1 that records her initial speed as she bursts upward.

This argument may be repeated and repeated to the same effect. A doubled antidifferentiation takes acceleration to position. The black arts having been celebrated twice, $P(t) = 1/2 \, gt^2 + C_1t + C_2$. But at $t = 0$, $P(0)$ is just C_2 itself, the *position* of a falling object before it has begun to fall. C_2 thus has a simple interpretation as the height of old—**H**, to recover a symbol.

It is a fact that the acceleration of a freely falling body is constant. A single and a simple fact has by the mathematician's potent arts been subordinated to a symbolic form capable of controlling *every* circumstance in which an object falls toward the center of the earth with a specified initial velocity and from a specified initial height, the symbolic form holding in balance acceleration, velocity, and position in a subtle miracle of coordination.

A Surprise Quiz

What lies between the brute *fact* that acceleration is constant and the general *law* of falling bodies?

The answer please? *Ah,* no one knows, perhaps because again I have succumbed to that painful pedagogical vice of asking questions to which I alone know the answers; but I *do* know the answer and the answer is worth knowing. There is *nothing* that stands between the fact and the law except the double operation of antidifferentiation, an operation that arises within and that is entirely defined by the calculus. It is here, as in so many other places, that an austere set of rules, a strange collection of definitions, and an elegant symbolic notation come together to permit an inescapable series of inferential steps leading inexorably from the interesting but undeveloped fact that gravitational acceleration is a constant to a law of nature, a form of words exerting its control over space and time.

One Last Thing

The elementary functions may all be differentiated by elementary rules, the process yielding elementary functions in turn. Antidifferentiation succeeds *by definition* for the derivatives of each of the elementary functions. This circle is closed. But given an elementary function, does it follow that its antiderivative must be elementary in turn?

No, unfortunately. There are plenty of elementary functions that fail of antidifferentiation if the standard of success is that the revealed antiderivative be an elementary function. The functions

$$\frac{e^t}{t}, \quad \frac{\sin t}{t}, \quad \frac{1}{\ln t},$$

and many others resist antidifferentiation in elementary terms. There is *no* elementary function $f(t)$ whose derivative is $\sin t/t$. This is not a mystery; it is just the way things are. A circle closed in one respect—

differentiation—is open in another—antidifferentiation. The hope that antidifferentiation might return an elementary function to an elementary function is a part of a system of childish illusions that dominates elementary mathematics; it is here, when this fact about antidifferentiation is grasped, that like first love elementary mathematics passes into memory.

Sermonette for Mr. Waldburger

Just as there are algorithms that make differentiation a quasi-mechanical process, there are rules that do the same for integration. In fact, certain techniques of integration are derived from techniques for differentiation. The chain rule provides an example.[4] As an algorithm for differentiation, it says that

$$\frac{dF}{dt} = \frac{dF(g(t))}{dg(t)} \frac{dg(t)}{dt}.$$

There is no reason why an indefinite integral cannot be stuck on both sides of this equation:

$$\int \frac{dF}{dt} = \int \frac{dF(g(t))}{dg(t)} \frac{dg(t)}{dt}.$$

The antiderivative of dF/dt is simply the original function F with a constant lumbering in its narrow wake. But properly speaking, F is a function of $g(t)$, and so function and lumbering constant have the form $F(g(t)) + C$, from which it follows that

---

[4] An example discussed in chapter 4.

$$\int \frac{dF(g(t))}{dg(t)} \frac{dg(t)}{dt} = F(g(t)) + C.$$

And despite its rebarbative appearance, this formula leads to a rule for integration, one expressed as a little theorem. Say (as I have said but said implicitly) that f and g are functions such that $f(g)$ and the derivative of g are all continuous on an interval I. I am now treating the derivative of g as a function in its own right. Now if F is an antiderivative of f on I, it follows from the formula just given that

$$\int f(g(t)) \frac{dg(t)}{dt} = F(g(t)) + C.$$

Here I have substituted the unadorned $f(g(t))$ for the derivative of F on the grounds that F is an antiderivative of f on I.

This little formula is known as the substitution rule of integration. It makes it possible on occasion to find an antiderivative of an otherwise inscrutable function—$f(t) = (t^2 + 1)^2(2t)$, for example.

A tomblike silence envelops the room. "Can *anyone* specify this antiderivative:

$$\int (t^2 + 1)^2 (2t)?"$$

Mr. Waldburger is prepared to moan dramatically, his habitual sign of intellectual respect, but moaning does not commend itself here as a strategy of discovery, and so the question I am asking has a cultural as well as a mathematical meaning. It illustrates the need for a technique in the calculus, a way of finding things out.

"It is the substitution rule that takes care of this case," I assure my already restless students. I rap the blackboard with my knuckle. "A few *expeditious* substitutions are all that is needed."

"*Yeah, yeah*," Mr. Waldburger says, depressed now by the chalky hen scratches that have gone up on the board.

"But no, really, a few expeditious substitutions *are* all that is needed. In particular, let the function $(t^2 + 1)$ be represented by $g(t)$. Then the derivative of g automatically represents the function $2t$—"

At this, Mr. Waldburger allows his moan to deepen.

"Because $2t$ is derivative of $(t^2 + 1)$ and $g(t)$, *it* represent $(t^2 + 1)$," Hafez the Intelligent breaks in brutally. "You don't eat so many donuts maybe you understand."

Lately, Mr. Waldburger has taken to bringing a carton of jelly donuts to class in order to console himself against the calculus.

"Then again," I say, "if $g(t)$ represents $(t^2 + 1)$, let $f(g(t))$ represent $(g(t))^2$, so that $f(g(t))$ is simply another way of saying $(t^2 + 1)^2$."

I pause to survey the effects that these remarks have on my class. Mr. Waldburger is endeavoring to more or less *hide* his jelly donut by cupping it with his palm. From time to time, he chews silently.

"With these substitutions in mind, it appears that

$$\int (t^2 + 1)^2 (2t)$$

is an *instance* of the formula

$$\int f(g(t)) \frac{dg(t)}{dt},$$

—*right?*—so that

$$\int (t^2 + 1)^2 (2t) = F(g(t)) + C,$$

where F is any antiderivative of f."

Is everyone with me? I have no idea why I ask: *no one* is with me.

"But it is easy to see," I go on—

"For you, maybe," Ms. Klubsmond says, convinced once again that the very chalk dust on the blackboard has engaged in a conspiracy against her.

"Easy to see that $F(g(t))$ equal $1/3(g(t))^3$ is antiderivative of f," says Hafez the Intelligent, "because derivative of $1/3(g(t))^3$ is $g(t)^2$ and f is function of form $f(g(t)) = g(t)^2$."

"Quite right, Hafez," I say. "So in terms of the original problem, the answer to my question is:

$$\int (t^2 + 1)^2 \, 2t = \frac{\left(t^2 + 1\right)^3}{3} + C."$$

"*Whee*," says Mr. Waldburger, the remains of a jelly donut still glistening on his downy upper lip.

I do not hold any grudges. The substitution rule is an *acquired* taste.

Later that day, I showed Igor the problem set on integration by substitution that I prepared for my class. He looked over the contrived problems, solving each in the blink of an eye. I was grateful that Mr. Waldburger was spared the demonstration of his power.

"Too easy," Igor said. And then with a sinister chuckle, he went to the blackboard of the mathematicians' lounge and quickly wrote a few truly monstrous integrals, great dense complicated formulas. Each, I could see, was more or less designed to make an application of the standard rules impossible.

"I assign these to my class on an examination," I said, "half my students will fail."

For a long moment, Igor M stood there, a tiny, fierce little man.

"Is nofink," he finally said. "In Moscou, *ch*all fail."

chapter 21

Area

Now you must imagine Lyndon Johnson—of all people!—tramping about his God-awful Texas ranch, a crowd of reporters behind him, throwing his arms wide to encompass the dry, humped hills, the pecan trees, and the little flowing creek, barking expansively into the V formed by his outstretched arms that *this here is all mine*.

But what is the it that is all yours?

This peremptory question prompts the president to stop in his tracks, turn, and with his eyes narrowed above his large nose, fix his interlocutor with a question of his own: *You shittin' me, son?*

Stammering in the hot sun, the mathematician standing in that multitude explains that he is merely curious about the concept, the abstract thing betokened by the president's ungainly outstretched arms—arms that take in along the lines of their imaginary extension a region of the earth, a piece of territory, an expanse, the phrases *a region of the earth, a piece of territory, an expanse* all of them ways of conveying the abstract concept of *area*.

Hearing all this out, the president draws himself up to his full six feet and five inches and says: *You better not be shittin' me, son.* Without another word, he stalks off into the harsh sunlight.

The Mathematics of a Rainy Saturday

In another way and in another world, the graph of a positive continuous function $f(t)$ appears on a Cartesian coordinate system, suspending itself as if on stilts between two straight lines at the points a and b:

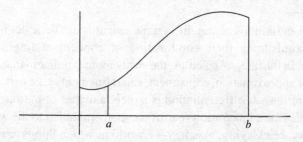

The curve is an active agent, a snake that by slithering between $f(a)$ and $f(b)$ acts to *enclose* a portion of space, a kind of territory. If it is territory—or call it what you will—that lies beneath the curve, *what is its area?* It is with this question, simple, yet divinely enigmatic, that the second part of the calculus commences. The question is simple because the shape of its answer has already been foretold: whatever the mathematician comes up with, it had better be a positive *number* that he assigns to all that *emptiness;* and just for a moment let me celebrate the audacity that drives the demand, the mad conviction that there should *be* a number that *could* express our entirely sensual appreciation of the fact that some *stuff* is spread out beneath the curve; but the question is also divinely enigmatic because it straddles the line between a peremptory request that a discovery be made—*look, go find the area*

underneath a curve—and an equally peremptory request that a definition be provided—*look, what do you mean by the area underneath a curve?*—thus suggesting yet again that in its innermost aspects mathematics is involved in bringing something into being by a choice of words.

Area is the second of the two chief concepts of the calculus and the second place where the real world reaches out to enforce a mathematical construction. As speed gives rise to the derivative of a real-valued function, area gives rise to its *definite integral*. Like a leaf folded along its axis, the calculus is divided into conceptual halves, derivatives on the one side, integrals on the other. But nothing in the development of the subject has yet justified the implicit symmetry that this image suggests. What has speed to do with area? It is a virtue of the calculus that the lines of its development are dramatic enough to provide an intimation of the answer, which, of course, is *everything*.

The definitions of the differential calculus are delicate; they are often exquisite in their combination of conceptual fragility and strength. In the integral calculus, the vastly more familiar mental movements of approximation, adjustment, and refinement come to dominate the definitions. If differentiation is largely a matter of putting butterflies on leashes, integration returns the mathematician to the world of carpentry, bricklaying, masonry—a world in which things are moved or made. But whatever the differences in their texture, differentiation and integration are alike in this crucial respect: the ideas in which they are ultimately embodied transcend their origins in order to achieve purely a formal existence.

Approximation is that most familiar of intellectual acts, the butcher, his thumb on one side of the balance, approximating the weight of the forbidden porterhouse on the other. It is to approximation that we repair when something whose properties are known is made to stand in for something whose properties are not. Area is under analysis, but the domain of objects whose area is already known is small and dominated almost entirely by a strangely limited series of plane figures: the square, the rectangle, the circle, a couple of chummy rhomboids, and not much else; it is surprising, now that the need arises, that Euclidean geometry draws the tent of its theorems over so sparse a setting, so curiously empty a stage. Nonetheless, the integral calculus looks back in gratitude toward the tried and trite formulas of Euclidean geometry. A formula is after all a form of words that *works.*

The area beneath a curve is not shapeless, but by the standards of Euclidean geometry it is disturbingly amorphous, its shape shifting somehow with small variations in the function generating the curve. It is all the more remarkable, then, that in a rough and ready way the area beneath a curve may be *approximated* by a succession of rectangles appearing on the plane like modernist skyscrapers:

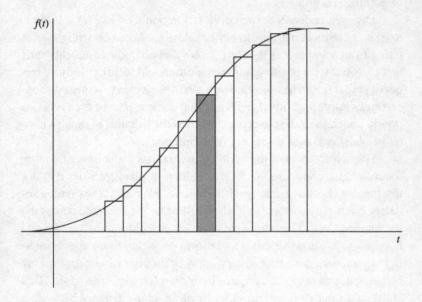

If this seems rather a crude approximation, it is useful to remember that it is valuable nonetheless by virtue of the fact that the area of a rectangle is something we know how to compute as the product of its base and its height.

The characteristic gesture of the integral calculus, in which an unexpected approximation is followed by a stretch back toward Euclidean geometry, is on display in all its bright pink innocence; it is the details that are yet lacking. The rectangles that have appeared thus far as illustrations need to be organized suitably, the approximation that then results, refined and refined again. The image of refinement being carried on and on suggests a mad painter endlessly adjusting an academic portrait, but it suggests as well that a *limit* may be in prospect, a place where refinement comes to a settled end. This suggestion Leibnitz could not hear: he had no recourse to limits and so followed his fertile imagination all the way to the end, each rectangle literally refined out of existence until it stood no wider than a straight line, one containing only an *infinitesimal* area. The sum of all those empty-area'd rectangles Leibnitz exhibited as the area beneath the curve, mindful, I am sure, that he was exhibiting a monster of sorts. The sum of nothing is nothing.

The area underneath the curve $f(t)$ is bounded by $f(t)$ and the t-axis; its outermost margins to either side are marked by straight lines rising from a and b. It is the stuff *within* that embodies area, this stuff that calls out for a positive number, a numerical identity. If these frequent verbal flourishes—*area, stuff, territory, expanse*—seem disconcertingly imprecise, this is evidence that some aspect of area remains poorly understood, the conceptual work of the calculus as much a matter of *clarification* as discovery or definition.

The construction of the definite integral has something of a rainy Saturday afternoon's remembered childhood pleasures: the radiator hissing, hot chocolate nearby with a bulbous marshmallow floating sedately on its gray surface; and a box full of Legos scattered across the linoleumed floor. The otherwise ordinary straight line between a and b—the t-axis—now becomes a platform on which those approximating skyscrapers raise themselves up. Any number of rectangles may be needed, and they each require their own foundational space. This childhood's pudgy finger provides by dividing the distance between b and a evenly by means of the n points of a *partition:* $a = t_0 < t_1 < t_2 < \ldots < t_{n-1} < t_n = b.$[1] These points separate the distance between a and b into evenly spaced intervals. The points go *from a* to b, and so a is identified with the first point and b with the last. Starting at a, each point that follows is to the left of the point that succeeds it, with t_1 less than t_2, and so on.

---

[1] It is sometimes convenient to begin a sequence with 0 as the first index, rather than 1. There is no hidden significance to the notation.

The space between *a* and *b* is an interval along a line; partition points segregate this interval into subintervals. The first subinterval is simply the interval from $a = t_0$ to t_1, the second, the interval from t_1 to t_2. The complete collection of subintervals the mathematician symbolizes in the obvious way as $[t_0, t_1], [t_1, t_2], \ldots, [t_{n-1}, t_n]$:

Picture and symbols say the same, obvious thing, that the length of line between *a* and *b* has been evenly chopped up.

The division of the line between *a* and *b* into *n* subintervals is a variable activity, the number of subintervals increasing as *n* itself increases, so that the imperious architect of this affair can command the line to divide into one or two or three or ten thousand parts. Whatever the number of subintervals, the *length* of each is symbolized by Δ, with subscripts indicating in the obvious way the length of the first subinterval (Δ_1), and then of the second (Δ_2), and of the third after that (Δ_3).

What *is* the length of each subinterval? Not to worry. The length of the first subinterval $[t_0, t_1]$ is a measure of the distance between two numbers t_0 and t_1. And distance on the line collapses simply into difference, so that the length of the subinterval $[t_0, t_1]$ is simply the numerical difference between t_1 and t_0. Now my assumption throughout is that the partition is *uniform*: each subinterval is of equal length. This makes for a convenient numerical measure. The length of any given subinterval is the distance between *a* and *b* divided by the number of intervals. This is $(b - a)/n$.

The construction of a partition completes half the work of organization; what remains to be accomplished is the actual *construction* of the approximating rectangles. The Lego box, with its vivid assurance on the cover that the construction of a skyscraper, dinosaur, or floating bridge requires only a few moments of work, is now consulted for the discreet directions on the side, directions that for any interesting project end, I seem to remember, in precisely the same way: *may require adult supervision*. Here, too.

Each subinterval, it says in step one, *serves as the base on which a rectangle is erected.* Step two follows obligingly: *The height of each rectangle is fixed by the value of the function f itself.*

The value of the function *f*? Let me see those directions again. The value of the function *f where*? It doesn't say.

It *needn't* say, meaning that *f* may take its values *anywhere* within a given interval.

The area of the first rectangle, step three informs me, is simply the product of its base and its height, presented as a retrospective gift from a Euclidean world. But both of these are known quantities, as step four makes clear, the base, Δ_1, and the height, $f(t)$, so that the area of the first rectangle is simply $f(t) \times \Delta_1$. What holds for the first rectangle holds as well for the second and the third; it holds for *all* of the rectangles for no better reason than the fact that they *are* rectangles.

There remains only a final step to navigate. Rising out of the ground, these Euclidean rectangles are designed to approximate the area underneath the curve. What does the approximating is the collection of those rectangles, an inner voice of balance saying that the area beneath the curve is approximately the same as the area of those rectangles taken as a whole, or smooshed together, or otherwise amalgamated. The simplest of all operations does the joining. The area of the rectangles taken as a whole is the *sum* of their individual areas. If there are *n* rectangles in all, their collective area is

$$f(t_1)\Delta_1 + f(t_2)\Delta_2 + \ldots + f(t_n)\Delta_n.$$

Such is a *Riemann sum,* the Lego box now fusing this childish construction into the dreams of the great Bernhardt Riemann, his dream becoming concrete.

A final detail. One symbol can do the work of many. Such a symbol is provided by the Greek letter *sigma,* a term signifying a sum, with

$$\sum_{i=1}^{n} f(t_i)\Delta_i$$

conveying the same information precisely as

$$f(t_1)\Delta_1 + f(t_2)\Delta_2 + \ldots + f(t_n)\Delta_n.$$

The variable i beneath the sigma is an *index* running from 1 to 2 to 3 to that last lingering n; it acts stroboscopically, changing values in a flash as each new term arises for inspection, the notation conveying the idea that at $i = 1, f(t_1)\Delta_1$ is up for inspection, and as i changes to 2, $f(t_2)\Delta_2$ comes up for review, with each inspected term added to the ones that have gone before.

Symbols now crowd the field, and even though they *cannot* convey more than can be conveyed in ordinary English, they begin to suggest that in some sense their proliferation upon the printed page is a sign of coarse vitality.

chapter 22

Those
Legos
Vanish

IN 1854, GEORG FRIEDRICH BERNHARDT RIEMANN OBTAINED A POSITION
something like that of an assistant professor at Göttingen; he was required to deliver an introductory lecture to the mathematics faculty.
He was but twenty-seven; he had twelve years to live.

The lecture was delivered in a large room, wooden floors echoing
as mathematicians and physicists enter the hall, dust motes dancing in
the heavy air, the thin sunshine slanting into the room through high
clerestory windows. The room finally fills, and there in the middle of
the hall is the great Gauss himself, an austere old man sitting erect, his
head high, the stiff linen of his white shirt rising to the lobes of his old
man's hairy ears; his hooded eyes are heavy. The other men fidget for
a time on the uncomfortable wooden benches. Someone coughs.

Dressed in a heavy black frock coat and shapeless black pants, Riemann rises from his seat in the front row and advances toward the lectern, almost tripping as he mounts the stairs leading to the stage; his upper lip is wet. Pushing his spectacles to the bridge of his nose with his index finger, he places the sheaf of papers containing his lecture on the podium and then, with an evident tremor in his thin tenor voice, begins to speak, his eyes fixed on the papers in front of him. He proposes, he says, to address the hypotheses that lie at the foundations of geometry. He coughs slightly and talks again; he talks carefully but without hesitation; he talks for the better part of an hour. And as he talks, there is a dawning sense among the members of the audience that by means of these calm and measured sentences, this slight, somewhat tubby young man is exploring an entirely new intellectual prospect, one of scope and grandeur, a geometrical world in which the last lingering connection to Euclidean geometry has been severed. When Riemann has quite finished, the great clock of intellectual discovery, which at all times is set to either before or after, stands now at after. Gauss rises from his seat, hardly looking at the men next to him, pauses in the aisle to adjust his shirt collar, and walks solemnly toward the podium to congratulate Riemann, the handshake marking one of those queer moments in intellectual history in which the series of brilliant gestures that constitutes the life of the mind is somehow completed *and* continued.

An Infinite Accretion of Detail

The work of approximation is now approximately complete. The region underneath the curve acquires its arithmetic identity by proxy, taking as a measure of its area the first available Riemann sum. But like the final product of an afternoon's construction, which despite the brimming pictures on the cover looks *nothing* like a skyscraper or a *Tyrannosaurus rex*, the approximation to the area underneath a curve thus far engendered seems painfully crude. No wonder. There is nothing in any of this that is new, nothing in fact that goes beyond that conceptual reach of the ancient Greeks, the notation excepted.

In the integral calculus, approximation is followed by refinement. It is in the very nature of approximation that it may be made better and better, the artist's circle passing to a portrait in steps, by means of an accretion of detail; but the principles that guide the artist's hand as he transforms a geometric shape into the slightly smiling face of a young woman, *these* no one knows, each artist learning the secrets in silence. Mathematics is simpler if only because its principles are more explicit, the mystery the more troubling because it is so often out in the open. In the case of the area underneath the curve, refinement proceeds by means of the simple principle that *as the number of rectangles increases, the approximation gets better and better.*

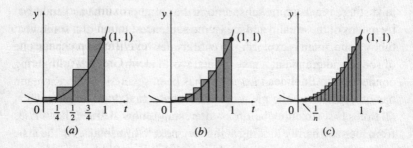

An increase in the number of rectangles is enforced by allowing n to become larger and larger. This has the effect of shrinking the width of each of the subintervals Δ_i. The principle in question may equally be conveyed saying that *as the width of each subinterval decreases, the approximation gets better and better.* Whatever the formulation, the principle should evoke little surprise. It is suggested by common sense, which observes that by allowing the rectangles to multiply, the degree to which they lie above or below the curve is diminished; it is suggested, as well, by the picture, and the principle itself was within the competence of the Greeks, who computed area by means of a method of exhaustion not markedly different from the scheme that I have set out.

It is not often that even in retrospect the very place where something is about to change may vividly be discerned, the old falling away, the new about to come into existence; but in the development of the integral calculus, this is the point, the very place. The requisite conceptual innovation is brought about as the imagination recognizes ruefully that no matter how the approximation is extended, the simple process of adjusting rectangles so that they come closer and closer to filling in space below a curve is hopelessly inadequate if only because hopelessly open-ended. No matter how fine the refinement, it may always be made finer as Δ_i gets smaller and smaller, small without end, but never small without width.

The language in which refinement is cast calls to mind an irresistible process, those widths shrinking steadily, rushing toward an inexorable end, crushed entirely out of existence. This way of putting things evokes the chief instrumental concept of the calculus, the nature of a process getting closer and closer to an inaccessible something suggesting, of course, that the aim and end of approximation might be revealed in a mathematical *limit*. With the introduction of a limit, the new steps forward to revivify the old, and by revivification changes it entirely, as the ancient, rich, intuitively evident Greek world steps backward into the shadows.

A limit, remember, is a fixed something toward which other somethings are tending. In the integral calculus, it is the Reimann sums that do the tending, and they do it as n gets larger and larger, as $n \to \infty$, in fact. The limit of Reimann sums, should it exist, is expressed by the formula:

$$\lim_{n\to\infty} \sum_{i=1}^{n} f(t_i)\Delta_i,$$

the elegant and compact symbols denoting the real and finite number toward which a sequence of finite sums is tending.[1] Note the double dip in this formulation: *a real and finite number,* and a *sequence of finite sums.* Once again, the invocation of a limit seems by an appeal to the infinite to skirt the margins of the ineffable, but when the symbols are studied soberly, everything, it would appear, is miraculously transmuted into what is fixed and finite. A limit is, after all, a number like any other; and at any given moment, a Riemann sum denotes nothing

---

[1] It is helpful, given the strangeness of the sigma notation, to remember that the limit in question is precisely the limit of the otherwise prosaic sums $f(t_1) \Delta_1 + f(t_2) \Delta_2 + \ldots + f(t_n) \Delta_n$ as n becomes larger and larger.

more than the collected area of a finite set of rectangles. This, too, is something measured by a fixed and finite number. Allow time to advance, and the rectangles multiply, but only by a finite amount, the Riemann sum that results nothing more than the collected area of a somewhat more numerous set of rectangles. And this, too, is measured by a fixed and finite number. It is only when the mathematical alchemist demands that this process be allowed to proceed as the number of partition points grows toward infinity that the shadow of the infinite falls across the experiment; but no sooner does the shadow appear than it is wiped away by the appearance of the limit itself, so that like some deeply hidden secret the infinitary magic remains buried beneath a finite surface.

Approximation is now contingent on the *existence* of those limits. Remarkably enough, it is a fact that if f is continuous and positive on $[a, b]$, it follows that the requisite limit exists. It is this circumstance that motivates the *definition* of area, that sustains the construction of a concept. A continuous and positive f given and on hand, the area bounded by the graph of f, the t-axis itself, and the lines rising from a and b is determined by an approximating area taken to the limit:

$$\text{AREA} = \lim_{n \to \infty} \sum_{i=1}^{n} f(t_i) \Delta_i,$$

where t is *any* point in Δ_i and Δ_i is equal in each case to $(b - a)/n$.

Area thus emerges from these considerations as a brand-new concept, a mathematical *creation*, the limit of a series of sums; and as such it is expressed in the calculus by a new symbol, one arising from a deliberate attenuation of the sign for a sum, the elegant *definite integral*

$$\int_{a}^{b} f = \lim_{n \to \infty} \sum_{i=1}^{n} f(t_i) \Delta_i$$

of a real-valued function f between the limits a and b.

The Integral Itself

The definite integral is the second of the new concepts of the calculus; the derivative of a real-valued function, the first. Like the derivative, the integral of a function issues in a number, but its conceptual structure indicates an implicit affinity for the concept of average rather than instantaneous speed. The definite integral does not appear to yield a number that may with ease be expressed as the range of a function. Beyond that, the definite integral stands opposed to the derivative in a profound and suggestive way. The derivative of a real-valued function is intensely *local,* the derivative alive and quivering at a point and a neighborhood of that point. The number signified by the definite integral is, by contrast, meant to characterize an entire region of the plane. The definite integral is a *global* concept, something that transcends entirely the concerns of the here and now, the this and then.

As its name might suggest, the definite integral has a no-nonsense quality somewhat at odds with its definition by means of a limit; its down-to-earth appearance owes much to the fact that the definite integral is a number and so serves the purpose for which it was intended, which was to extend the ancient program of replacing geometrical aspects of the world—distance and area—with uniquely pregnant tags, symbols drawn from an alien numerical world. But if the number has a reassuring palpability, it is nonetheless a number reached by means of a limiting process, and so something new in intellectual experience. Between the area of a parallelogram in the Euclidean plane and the area underneath a curve, there is a common concept, that of area, but absolutely no common method, and so the two numbers that result are representatives of two different visions.

The definition of the derivative, and so the concept of instantaneous speed, is accomplished through the sleight-of-hand that is an inexpungeable part of any passage to the limit. A series of discrete operations is invoked and then allowed significantly to accumulate, as if the passage to the limit were rather like one of those nineteenth-century movies in which a series of pale repetitive scraps would be flipped faster and faster until at a single moment the viewer had the eerie experience of seeing the discrete—those scraps, taken one after another—merge and become continuous, showing to the astonished eye a plump Victorian bather shyly disengaging from her formidable corset. The same process is at hand in the definition of the definite integral. Those imagined rectangles go up, all in a row; their area is taken; those areas are summed. All this is discrete as discrete can be; a pat series of patterns, hard-edged as the *ka-chunk* of those slides (or whatever) being introduced into that nineteenth-century movie projector; but now, pass to the limit and those rectangles begin to blur, their bases becoming narrower and narrower, the approximation to the curve becoming ever finer as the poor square things begin to touch the swaying curve, and *there* in the limit itself those rectangles reach up to touch the curve at a point of perfect coincidence.

For all that, the definite integral yet expresses a number. As a result, area undergoes an inevitable contraction, the quantity (or quality) vanishing in favor of the number that describes it, the whole of something, all that space, the Pedernales, those pecan trees, canceled in favor of a symbol, the lush particularity of this place, these trees, that swollen creek, demoted and then dismissed.

But this is true of any attempt to describe the world, which remains forever more various than the symbols used to contain it.

The Integral Wishes for a Mean Value

The light from the mean value theorem flickers over the calculus; it flickers over all of elementary mathematics, and in view of its importance it is natural to wonder whether something like the theorem holds for the definite integral.

It does. The result is a theorem with an unusually obvious conceptual structure. Seen in just the right light, the mean value theorem for integrals is an immediate consequence of the mean value theorem itself. Let f be a continuous and positive function on $[a, b]$ and let M and m stand for the maximum and minimum values of f over this interval. Continuous as it is, f takes both these values.

Two rectangles now come into play. The upper rectangle is formed as the product of M and $b - a$, and the lower as the product of m and

$b - a$; these rectangles either contain the area underneath the curve $f(t)$ or they are contained within the area underneath the curve $f(t)$.

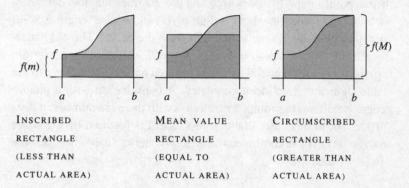

INSCRIBED	MEAN VALUE	CIRCUMSCRIBED
RECTANGLE	RECTANGLE	RECTANGLE
(LESS THAN	(EQUAL TO	(GREATER THAN
ACTUAL AREA)	ACTUAL AREA)	ACTUAL AREA)

The area underneath the curve is itself expressed by the definite integral from a to b:

$$\text{AREA} = \int_a^b f,$$

and all this is simply a matter of rehearsing a few familiar concepts.

Two facts now enter into discussion without proof. Whatever the area underneath the curve, it is *at least* as large as the area of the lower rectangle and *no larger* than the area of the upper rectangle, those upper and lower rectangles serving as the gray and obdurate margins beyond which the area underneath the curve cannot go and will not proceed.

Evident from the pictures, this, too, is evident from the very definition of M and m and the definite integral.

Now the definite integral is the supreme expression in mathematics of approximation taken to the limit, its contribution to the calculus a decisive demonstration that approximation may be continued to the point that it becomes perfect coincidence. And yet the approximations proceed from something so simple as a series of rectangles, ancient, Euclidean figures improbably associating themselves with calculations concerning curves. It is natural in this context to wonder whether the area of *some* rectangle might express the area underneath the curve perfectly.

Not any rectangle, of course, for in that case the question is trivial and answered trivially. The question is altogether more specific. Is there a point s in $[a, b]$, the poised and precise question now demands, such that the rectangle whose height is $f(s)$ and whose length is $b - a$ coincides in area with the area underneath the curve? The old haunting refrain: *Is there something? Does it suffice?* Question and refrain turn the imagination backward from the area underneath a curve to the stable figures of Euclidean geometry. A consultation with a picture generally suffices to prompt a careless assent: there *is* a number; it *does* suffice; but in this case, that careless assent is justified by the sober analytic facts. If f is continuous on a closed interval $[a, b]$, there *is* some s in $[a, b]$ such that

$$\int_a^b f = f(s)(b-a).$$

Such is the mean value theorem for integrals; its proof is lighthearted, one quicksilver step after another resembling in their effect the cascade of a cold mountain stream.[2]

In the Stream of Time

Newton and Leibnitz understood the need for the definite integral, and the notation now in use was designed by Leibnitz, a man of great notational inventiveness; but Leibnitz thought of integration as he thought of differentiation—in terms that were logically incoherent. The infinitesimal quantities that appeared in his definition of the derivative appear again in his definition of the integral, so that finding the area beneath a curve involves summing infinitely many infinitely small strips. You must consider the especially delicious head shakings and hand waggings that attended a demonstration by Leibnitz of just how it was that infinitely many things without area could by an ineffable mechanism come to a finite positive sum.

But Newton and Leibnitz were men of genius, and that they had

---

[2] For the quicksilver proof, see appendix 2.

recourse to what now appears nonsense is evidence of how long the road ahead of them was bound to be. Not that Euler or Lagrange or Monge or Maupertuis did much better. The great glittering circle that appears in French culture between the end of the seventeenth and the end of the eighteenth centuries had beside its brilliance this characteristic: it was made up of men who for the last time in the history of mathematics felt entitled to say with a shrug when asked precisely what they meant: *Ah, but gentlemen, we have no idea.*

Sitting in a dark, wood-paneled room in what is now the sixteenth arrondissement, it was the busy, indefatigable Cauchy who formulated the modern definition of the integral in the early part of the nineteenth century, and Cauchy, with his superb organizational powers, who saw what needed to be done and did it; but in Cauchy's development of the integral a crucial step, astonishingly enough, is simply missing. Cauchy lacked a concept—*uniform continuity,* as it happens. A great mathematical idea is often created in stages, and if it fell to Cauchy to purge the concept of area of its metaphysical messiness, to give it an essentially modern shape, it fell to Riemann to endow the concept with its full measure of mathematical formality.

Born in 1826, in a village near Hanover, Riemann appears in retrospect as one of the nineteenth century's hopelessly poignant figures, akin to Schubert in his strange and compelling talent and akin to Schubert again in his horribly bad luck, Schubert having died young, Riemann dying younger yet, calmly coughing away his last minutes beneath a fig tree in Italy.

In his mathematical education, Riemann was influenced by the high mathematical culture of Germany; he was, after all, a student of the great Gauss, the old man coming to praise his Ph.D. dissertation in terms of almost unprecedented warmth. He was also influenced at a distance by the brilliant scattered lights of eighteenth-century French mathematical culture: Legendre, Lagrange, Euler, Monge all whispering to him through their books and papers in their malicious and captivating way. If Gauss offered the example of sublime mathematics undertaken as the intellect explores itself and nothing else, the French offered Riemann the example of mathematics as a fascinating game, one played on a sparkling board, the moves like silver seen in moonlight.

But Riemann was not like the French, nor wholly a pupil of Gauss. He was in his temperament a geometer, in his affiliations a Platonist, in his soul a visionary; he saw through appearances to a world less voluptuous and less complex than the real world, but more ordered, harmonious, stable, and beautiful. Many mathematicians, it is true, are Platonists, and most think of themselves as visionary. Riemann remains unique. In everything he touched, he brought strange and luminous gifts. He gave to complex function theory, that most exquisite of disciplines, one of its enduring forms; he withdrew geometry from its ancient Euclidean harbor; he conjectured that the physical properties of space would turn out to be aspects of its curvature. He formulated the concept of Riemannian manifolds; he attended to the distribution of prime numbers. He was interested in theoretical physics and even in psychology. Yet none of this quite captures the man. Alone among the mathematicians of the nineteenth century, he saw what he needed to see before ever he acquired the symbolic apparatus with which to express his vision; his certainty about each of his discoveries was richly merited, but exotic and spooky. Other mathematicians appreciated his accomplishments, but he remained alone, isolated by the *singularity* of his imagination. His vision of a physics enabled by geometry is one that attracted the young Einstein, who with the candor that only genius confers acknowledged Riemann as his predecessor. A strain of prophetic insight runs through their work, a way of investing sight with the aspects of a symbolic form so that the nature of things comes to be discerned simply by the act of seeing them.

Like Saint Bernard or Don Quixote, Riemann has entered into the mythology of the West. Round-faced, bespectacled, painfully shy, afflicted with a stammer, Riemann is nonetheless an infinitely moving figure, at once decent, kind, docile, affectionate, and gentle; despite his death, mathematicians see him advancing through the years in his tentative sleepwalking way, a small smile playing over his puffy lips; with his brown eyes opened wide and staring, at any moment he appears perplexed until seeing something that no one else can see he stops, murmurs *noch so was* in a tone of voice suggesting appreciation more than anything else, and then, his eyes alive and liquid and still kind, he smiles and moves on.

The Integral Wishes
for a More
Formal Existence

The definite integral defined simply to express the area underneath a curve is really an engineer's creation; it is too simple in its structure, too readily accessible, to be entirely satisfying to mathematicians like us. Let me rehearse the obvious objections: the function f undergoing integration must be positive; it must be continuous, and the partition over which the integral is defined must be uniform, its subintervals the same size. Some sense of generality scruples at these restrictions. The integral wishes for a more formal existence.

The restrictions I have just listed may be relaxed, and as they are, a more formal, an entirely more dignified, mathematical object comes into existence. The function f is now emptied of content to the point of anonymity, existing only insofar as it is defined over the closed interval $[a, b]$, where it conveys arguments to their values without ever exhibiting precisely how the conveyance is accomplished. The distance between b and a is divided, as before, into subintervals with widths Δ_i, but no particular effort is made to see that they are all equal; the point t_i is allowed to be *any* point whatsoever in the ith of the intervals.

The expression

$$\sum_{i=1}^{n} f(t_i)\Delta_i$$

denotes as before a Riemann sum for f, and if such sums are familiar, they are now also darkly enigmatic: f is, after all, anonymous and the various partitions of $[a, b]$ are not all of them the same in width.

The essential idea in the definition of the engineer's definite integral is *refinement,* partitions getting ever finer, the number n of rectangles increasing without limit. The engineer trusts in refinement because the partitions he envisages all break an interval into equal subintervals. But that demand has been relaxed in the interests of generality, and letting the *number* of subintervals increase without limit does not by itself ensure that the subintervals will *all* undergo an appropriate contraction, that some of them will do anything other than sit there stupidly.[3]

The appropriate effect may be achieved by demanding that the width of the *largest* subinterval go to 0. Let the width of that subinterval be denoted by $\|\Delta\|$; the Riemann integral may now be defined as the limit of Riemann sums as $\|\Delta\|$ shrinks into nothingness:

$$\lim_{\|\Delta\| \to 0} \sum_{i=1}^{n} f(t_i)\Delta_i = \int_a^b f.$$

With that integral entered into existence, f is said to be *integrable* over $[a, b]$.

The Riemann integral is the grizzled professional's definite integral, the grizzled professional's concept. Its relationship to the concept of area is now formal and conceived at a great distance. If f is positive and continuous, then *this* integral, like any other, represents the area underneath the curve between the limits of integration. But nothing requires that f be continuous and positive, and what the integral is doing on those other occasions of its use is now a matter between the integral and its conscience.

The great merit of the Riemann integral lies in its generality; but as with everything in mathematics (or in life), advantages obtained along one dimension are paid for by disadvantages incurred along another. When the commitments made by the Riemann integral are made explicit, a simple, clear, and lucid concept becomes irksome in its complexity. Consider thus the meaning of the limit used in the definition of the Riemann integral—that for every $\epsilon > 0$ there exists a corresponding $\delta > 0$ such that for $\|\Delta\| < \delta$,

---

[3] Suppose, for example, that the interval between 0 and 1 is partitioned according to the following scheme:

$$0 < \frac{1}{2^n} < \frac{1}{2^{n-1}} < \ldots < \frac{1}{8} < \frac{1}{4} < \frac{1}{2} < 1.$$

The largest subinterval in this partition has width 1/2, and its width remains 1/2 no matter how n is increased.

$$\left| \int_a^b f - \sum_{i=1}^n f(t_i)\Delta_i \right| < \epsilon,$$

meaning in turn that the definite integral may be approximated to any degree of accuracy by a Riemann sum.

For once, the vernacular is clearer than the formalism it is meant to explain, for the mathematics obscures a double dependency. Riemann sums are sensitive to the choice of t_i within the ith subinterval. It matters to the formation of the sum whether f is evaluated at its maximum in a subinterval, its minimum, or someplace in between. And yet if the definition of the definite integral is to make sense, this dependency must disappear, the limit arising *regardless* of the choice of t_i within the ith subinterval.

By the same token, the approximation signified by the inequality is sensitive to the partition as well as to the choice of t_i within a given subinterval. The definition of a limit does not demand that the approximation succeed regardless of the partition; only that the approximation succeed whenever refinement of a partition has proceeded apace. It is for this reason that the definition appeals only to partitions whose *norm* $\|\Delta\|$ is less than δ.

This double dependency makes it technically difficult to work with the Riemann integral.

Mean
Values
for Integrals

A continuous function f on an interval $[a, b]$ takes a maximum M and a minimum m, so that $f(t)$ always lies somewhere between these extremes.[1]

It is obvious that the integral of f over $[a, b]$ lies between these upper and lower rectangles:

$$m(b-a) \leq \int_a^b f \leq M(b-a).$$

Dividing by $b - a$ returns

$$m \leq \frac{1}{b-a} \int_a^b f \leq M.$$

The quicksilver point in the proof is now coming. The equation just given states that

$$\frac{1}{b-a} \int_a^b f$$

is some number between m and M.

---

[1] The Sign of Three, again.

But f is continuous and so somewhere *assumes* the values m and M. But an intermediate value theorem, making a special guest appearance here and nowhere else, steps in to remind us that a function f taking extreme values takes intermediate values as well. There is some point s, this theorem says, such that

$$f(s) = \frac{1}{b-a} \int_a^b f.$$

Now multiply both sides of this equation by $b - a$ and the result is

$$f(s)(b-a) = \int_a^b f;$$

but this is just what the mean value theorem for integrals affirms, the picture now justified by a few careless manipulations, the manipulations made vivid by the picture.

Piece of cake, as we in the business say; but perhaps a piece of cake as *only* we in the business say.

chapter 23

The Integral Wishes to Compute an Area

The integral enables you
To do what you need not do.
The theorem that will make this plain
Is one designed to spare you pain . . .

<small>MATHEMATICIAN'S DOGGEREL</small>

THE FUNCTION $f(t) = t^3$ ASCENDS FROM THE ORIGIN AND RISES UP THROUGH the space of a Cartesian coordinate system to intersect a vertical line moving upward from the t-axis at 1.

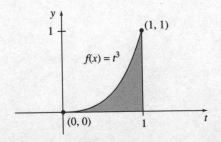

It is at the origin that this curve commences its activity and at the line $t = 1$ that it comes to exhaust its interest as an example; the expanse determined by the t-axis, the y-axis, that straight line, and the curve itself plainly forms something that should in the nature of things be assigned a number as its *area*.

Enter thus the cocky integral

$$\int_0^1 f,$$

ready to be vested with meaning. The vesting proceeds by means first of the imposition of a partition, with the interval [0, 1] subdivided into n equal intervals, so that Δ_i is $1/n$. Since f is continuous and positive, f may be assessed anywhere within each subinterval, the assessing done actually at the point i/n.

This means that if [0, 1] is divided into three parts, $n = 3$, and $i/3$ equals in the first case, 1/3, in the second, 2/3, and in the last case, 3/3 or 1. The index i is simply allowed to take on the values 1, 2, and 3 successively, and each time it takes on a new value a new fraction results.

The definite integral describes the area underneath a curve; it describes thus the area underneath *this* curve; and it does the describing by means of certain sums taken to their limit:

$$\lim_{n \to \infty} \sum_{i=1}^{n} f(t_i)\Delta_i.$$

But the shape that Δ_i takes has already been determined—Δ_i is equal to $1/n$. And the shape of $f(t)$ — that too has been fixed by the decision to assess f at the points $t_i = i/n$. The limit thus plays over a sum rewritten in terms of these identities:

$$\lim_{n \to \infty} \sum_{i=1}^{n} \left(\frac{i}{n}\right)^3 \left(\frac{1}{n}\right).$$

Now the product of $(i/n)^3$ and $1/n$ is equivalent to the product of $1/n^4$ and i^3 so that the formula just given may be rewritten as

$$\lim_{n \to \infty} \sum_{i=1}^{n} \left(\frac{1}{n^4} \right) (i^3)$$

by means of nothing more than the remembered rules for handling fractions.

How? The fraction $(i/n)^3$ is just $(i/n) \times (i/n) \times (i/n)$. Carry out those operations by multiplying the top by itself three times and the bottom by itself three times and the result is just i^3/n^3. Now multiply *that* by $1/n$ and you get i^3/n^4. But that is the same thing as $i^3 \times 1/n^4$.

You *see?* You *do* see? *Of course* you see.

This product in hand, a second reformulation of the symbols is in prospect when $1/n^4$ is moved in *front* of the summation sign:

$$\lim_{n \to \infty} \frac{1}{n^4} \sum_{i=1}^{n} i^3 .$$

Can I do that?

Absolutely. After all, $(2 \times 3) + (2 \times 4)$ may equally be conveyed by writing $2 \times (3 + 4)$, with the number 2 pulled from those factors and stuck in front. That is all that I am doing with $1/n^4$—pulling it from its factors, sticking it in front. In algebra, this is known as the old distribution trick, one of a clan of commonsensical laws.

There now follows a fact that no one knows, except us old calculus hands, that the sum of i^3 when i is first 1, and then 2, and then 3, and then ultimately n is itself given by a simple formula, a combinatorial charm:

$$\sum_{i=1}^{n} i^3 = \frac{n^2 (n+1)^2}{4} .$$

(The formula may be verified—but not demonstrated—by letting $n = 2$. Thus $1^3 + 2^3$ is 9, but then so again is $2^2 \times 3^2$ when divided by 4.)

Swap for the summation in

$$\lim_{n\to\infty} \frac{1}{n^4} \sum_{i=1}^{n} i^3$$

the expression just derived, $n^2(n+1)^2/4$, and the result is

$$\lim_{n\to\infty} \frac{1}{n^4}\left[\frac{n^2(n+1)^2}{4}\right].$$

A penultimate reformulation now follows:

$$\lim_{n\to\infty} \frac{n^2(n+1)^2}{4n^4},$$

the new formula coming about as $1/n^4$ is called upon to merge with $n^2(n+1)^2/4$.

This mathematical merger, I had better say at once, is only a matter of *multiplying* two fractions. The fraction's denominator is formed by multiplying n^4 by 4 and its numerator by multiplying $n^2(n+1)^2$ by 1.

A final reformulation follows, with

$$\lim_{n\to\infty} \frac{n^2(n+1)^2}{4n^4}$$

passing over to

$$\lim_{n\to\infty}\left(\frac{1}{4} + \frac{1}{2n} + \frac{1}{4n^2}\right),$$

this step proceeding by algebraic deconstruction. First, $n^2(n+1)^2$ *and* $4n^4$ are both divided by n^2. This leaves 1 in the numerator and $4n^2$ down below. Next the $n+1$ term is multiplied by itself, yielding

$$\frac{n^2+2n+1}{4n^2}.$$

But this fraction, when retrofitted to accommodate the addition in the numerator, is simply

$$\frac{n^2}{4n^2} + \frac{2n}{4n^2} + \frac{1}{4n^2},$$

and *this* fraction, when simplified, is simply

$$\frac{1}{4} + \frac{1}{2n} + \frac{1}{4n^2}.$$

The reformulation of formulas comes to an end with the recollection that the definite integral is equal to a certain quite definite limit:

$$\int_0^1 f = \lim_{n \to \infty} \left(\frac{1}{4} + \frac{1}{2n} + \frac{1}{4n^2} \right).$$

In order to come up with that limit and hence a number, it remains only to sift through the symbols in order to see where they are tending.

It is easier than it looks. As n gets ever larger, $1/2n$ and $1/4n^2$ both become progressively smaller; in time they dwindle into relative insignificance. This leaves $1/4$ as the only enduring expression at the limit, the only source of mathematical identity.

A limit having been reached and ratified, the definite integral takes on a definite value:

$$\int_0^1 f = \frac{1}{4}.$$

But the integral was conceived as a measure of area and so $1/4$ emerges from these considerations as the number signifying the expanse underneath the curve $f(t) = t^3$ between 0 and 1.

Now a secret must be imparted. This has been a very wearisome calculation. The suspicion might arise that the definite integral is a concept that bulks large but weighs little. The suspicion is correct to this extent: the definite integral having been defined, the definition does not determine how it is to be *applied*. But the technique of application the fundamental theorem of the calculus provides, and so what is missing in the definition appears later in the theorem, the play between the definition and the theorem a reminder that even in mathematics the formal divisions between a definition and a theorem may not reflect anything of importance about the world they are both intended to describe.

chapter 24

The Integral Wishes to Become a Function

LIKE A FOREIGN CITY SEEN AT NOON, THE CONCEPTS OF THE CALCULUS seem strange even when they are seen clearly. The explicitness that attends the mathematics is often of no help at all, and the connections that might serve to make sense of things are apt to be lost in the sun's glare. The Riemann integral, for example, suggests disconcertingly the appearance in the calculus of a fancy kind of formula, something like the Euclidean formulas for the area of a square or circle and differing from them only in the appeal that the Riemann integral makes to the concept of a limit. And yet there is *my* claim that area and speed dominate the large arena of the subject. Between the formula and the claim, however, meaning tends to vanish, the system of signs and shops by which one hopes to navigate around a strange series of streets impertinently refusing to arrange themselves in the desired and briefly seen pattern, the alleyway leading in memory to a sun-drenched piazza leading instead to an interminable series of inner courtyards, or a sullen square dominated by a squat and ugly church.

The great fact about the definite integral is simply that it is a number; and in the end, after the details and definitions have accumulated, the definite integral remains a number, tied through the circumstance of its identity to the most fundamental of mathematical objects. Instantaneous speed is similarly a number—the speed assigned by the mathematician to a point in time—but it is a number reached by means of an elaborate and far-flung set of definitions and constraints, and a number, moreover, that *also* figures in the calculus as an element in the range of a real-valued function. Although area and speed *are* the two crucial concepts of the calculus, they are in their development asymmetrical, answering to quite different imperatives. Area is in its essential nature more like distance than speed. It functions as a simple measure of elementary extent—*extent* because some quantity of stuff is being assessed, and *elementary* because the very act of endowing a line with arithmetical structure or space with the properties of a Cartesian coordinate system creates the intervals on the line or regions of the plane that *have* distance or that *possess* area.

And yet it is also true that in abstract matters the calculus is the party of the function, subordinating, whenever possible, all other considerations in its drive to express relationships, forge connections, tie together in a mathematical instrument the far-flung aspects of experience. Area is an absolute number, but attendant upon area there is also a function, one in which the old-fashioned claim that *this* region has an area measured by *that* number comes to be replaced by the altogether more sophisticated counterclaim that area is a quantity that can vary with time and that may in consequence *also* appear as the range of real-valued function.

The Grizzled Old Coach

Pacing by the sidelines, the grizzled old coach, a blue windbreaker over his still massive shoulders and a wedge of bubble gum moving rhythmically from one square jaw to the other, looks over the field where his team has taken possession of the ball on its own goal line. Looking very much as if his blood pressure were approaching transcendental levels, the grizzled old coach continues to pace the sidelines as his team moves the ball upfield, passing first, the completion prompting the coach to bang his fist into his palm, and then running, the tight end taking the ball and with an astonishing burst of speed outsprinting pursuers until he is driven out of bounds at the field's halfway mark, and then passing again, a perfect high, looping spiral caught in the end zone by the agile wide receiver, who seizes the ball from the air as it drifts lazily over his shoulders.

Afterward, in the locker room, sportscasters interview the grizzled old coach.

"Key to today's game, Coach?"

"Coming up that last drive, we controlled the field pretty good," the coach avers with satisfaction.

"Stick to the fundamentals?"

"Control the field, control the ball," says the coach.

Afterward, the sportcaster and his sidekick, a former athletic eminence, reverently go over the coach's remarks.

"So, Ed, the man's talking fundamentals. How's that square with what you saw down there today?"

"Well, Ted, the man, he's got the stats to back up what he says."

"He certainly does. But what do you think he meant talking about *controlling the field*?"

"Probably talking about letting area be a real-valued function of time there."

"You think?"

"Got to be, Ed. After all, the man, he knows the calculus."

If I draw the curtain of charity over this scene, it is only because Ed's succinct remarks are improbable, not because they are inaccurate, the grizzled old coach, taking his cue from military theorists, having in mind precisely a vision of the field in which his players by dominating play acquire domination over first a part of the field, and then successively larger and larger parts of the field, until in the end they control the entire field, dominating its *expanse* and bringing its *area* under their discipline and management; but what the coach sees in terms of mooselike men moving up and down a football field, the mathematician sees in more abstract terms yet, the football players and the football disappearing along with the crowds, the field, and the coach himself, until all that remains is the field imagined as a regular expanse, something that takes position on a Cartesian coordinate system as a rectangle lying flush along one coordinate axis. As the football game proceeds in real life, the mathematician's ghostly game reveals only the abstract essentials of play between time and area, progress to the twenty-yard line giving way to the area accumulated up to the twenty-yard line, with subsequent plays represented in the mathematician's game by a moving accumulation of area.

Time is represented as it has always been represented, but the space to which the mathematician now attends is localized as some spatial *expanse* or stuff, the product, in the case of the rectangular football field itself, of the field's base and height. The grizzled old coach paces the sidelines, dreaming, as all such grizzled coaches really do, of split formations and tight ends; far away, at least in spirit, the mathematician draws the finger of attention up the field, seeing as the result of simple algebraic operations, an abstract quantity such as area grow and swell.

The impulse to subordinate numbers to functions is in evidence in the creation of instantaneous speed, average speed disappearing simply because of its insusceptibility to functional expression. The same system of impulses at work in the case of average speed is at work in the case of area and the definite integral. If the definite integral measures the area underneath a curve, its associated function measures the area underneath a curve *up to some varying point*—up to *t* at first, then to some spot past *t*, and then to some spot past that, so that at each measured moment the definite integral issues a particular number, the string of numbers thus recorded forming the range of an associated real-valued function.

Starting then with the definite integral

$$\int_a^b f$$

of a function *f* defined on [*a*, *b*], a function *G* is created from the integral by allowing the upper limit of integration to vary

$$G(t) = \int_a^t f.$$

Note the small but significant notational change: the upper limit of integration is now set at the variable t and *not* the parameter b; integration still proceeds as it has proceeded, by means of an accretion of sums, but starting at a, the process of integration is now variable, $G(t)$ recording at each and every t the extent of integration up to that very point. Such is the *indefinite integral* of f, and although both definite and indefinite integrals are alike in being integrals, they are different in their most crucial respects. The definite integral denotes a specific number, something fixed, and as such the definite integral belongs to the world of things and their properties. The *in*definite integral is a function, mutable, changing with t, and as such the indefinite integral belongs to a world of things and their *relationships*. If the definite integral represents area, this integral measures something like *area up to a given point,* the awkward choice of words recalling the equally awkward *speed now*. And for the same reason, the awkwardness in both cases arising from the fact that ordinary English lacks a vocabulary in which to express the domination of the calculus by means of the concept of a function.

The indefinite integral now takes its place among the fundamental concepts of the calculus, the last link in a chain that needed to be forged, and with its definition, the forging is finished. But its introduction, I suspect, does not have the effect of establishing once and for all the connection among the concepts already in place. To what use is this new concept put?

The answer is clear and unexpected. What follows may serve as a coming attraction. The indefinite integral is above all important as the instrument by which new functions are created, the plain and prosaic extension of area into the indefinite integral serving in what is to come as the fecund source of creation, the place where the new is generated from the old.

In the cosmic catalog of elementary functions, few of the stalwarts were introduced by anything like a formal definition, trigonometric, logarithmic, and exponential functions appearing as admixtures of memory, intuition, and a few explanatory remarks. The indefinite integral makes for a way of *creating* these functions, bringing them into existence by means of a geometric charm.

The function $1/t$, considered at positive values of t, describes a graceful hyperbolic curve in a Cartesian coordinate system:

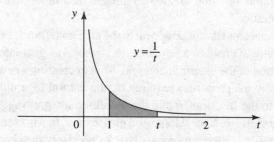

This function is continuous and so admits of integration between the limits of 1 and 2:

$$\int_1^2 \frac{1}{t}.$$

Since the function is continuous and positive, the integral offers an assessment of the area underneath the curve. With the assessment allowed to vary, the indefinite integral makes an appearance,

$$\int_1^t \frac{1}{t},$$

the result a brand-new function G, one taking its arguments from the interval between 1 and 2 and allowing its values to be determined by the values of the indefinite integral:

$$\int_1^t \frac{1}{t} = 6(t)$$

Remarkably enough, the function G defined by this process of indefinite integration has a clear, recognizable identity: it is, in fact, the natural logarithm, one of the elementary functions, a part of the cosmic catalog, so that

$$\ln t = \int_1^t \frac{1}{t}$$

follows as a *definition* of the logarithm for values of t greater than 0, an old, familiar function created by means of a new, daring mathematical operation.

This simple and dramatic appeal to the area underneath a curve succeeds not only in creating a new function—*that* was guaranteed by the very definition of the indefinite integral—but in creating a new function with precisely the properties required by the natural logarithm, so that the appeal to the function $1/t$ and the indefinite integral suggests more than anything else an actor slapping on grease paint in what seems a slapdash way only to emerge moments later as precisely the character in a Shakespearean play whose vivid features figure in the program notes.

If

$$\ln t = \int_1^t \frac{1}{t},$$

then **ln**1—the logarithm at 1—is bound to be

$$\ln 1 = \int_1^1 \frac{1}{t};$$

but the area underneath the curve between 1 and 1 is 0, and so ln 1 is 0 as well. This is precisely the property a logarithmic function must possess, and here it comes by its possession effortlessly.

There is more. At 1, the logarithmic function is 0. Elsewhere the logarithm ascends upward as a stately increasing and continuous function. Somewhere it crosses the line $y = 1$:

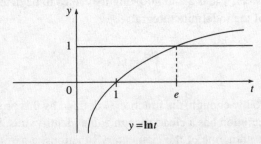

The black jewel of the calculus

Continuity comes to the fore here to guarantee (via a variant of the intermediate value theorem) that there is an argument at which the logarithm takes on this value, some place in its measured movement such that *there* the logarithm equals 1. The requisite number mathematicians denote by *e*—and *lo*, it is precisely the *e* of old that now enters into existence, the mysterious transcendental number figuring in the definition of the exponential function. Here it is not so much defined as *discovered,* the discovery more a matter of brushing away the debris so that the number stands revealed on the number line, an unexpected but familiar object recognized all at once for what it is.

The Integral Wishes to Offer a Consoling Word

It was from Alonzo Church that many years ago I took a course in mathematical logic for which I was unprepared—unprepared, that is, for the discipline of mathematics, unprepared for the demands of argument, and unprepared for Church's glacial and remote style. Church was an enormously distinguished mathematician. The material was very difficult, so difficult that someone had occasion once to *complain* about the complexity of a proof.

Church rotated his large torso away from the blackboard and toward the ten or so of us sitting in the lecture room. "*Any* idiot," he said calmly but with immense conviction, "can learn anything in mathematics. It requires only patience." He seemed curiously moved; a film came over his eyes. "Now to *create* something," he said, "that is another matter." In that queer moment of insight occasionally vouchsafed the very young, I understood instantly that Church was not reveling in his own accomplishments, but, with his own eyes fixed on the unattained goals to which *he* had aspired, was confessing obliquely to us, an audience of impossibly callow young men, that when it came to mathematics he, too, belonged in the company of humanity's idiots.

As do we all.

chapter 25

Between the Living and the Dead

NO LESS THAN ANY OF THE OTHER ARTS OF CIVILIZATION, A GREAT THE-
orem is an act of affiliation between the living and the dead. The simple statements and clear and precise definitions that *now* express its
meaning must always be appreciated as the product of historical fore-
shortening so that each affirmation is seen to carry with it some residue
of the years spent in anxious groping and fruitless search by men who
did not know how it would all turn out. Studying alone in the dark, with
only a desk lamp throwing a yellow light across the page, each reader
re-creates in his or her own experience the work of centuries. I am in
this chapter of the party of Newton and Leibnitz, of Euler and Lagrange
and Legendre, of Cauchy and Riemann and Dedekind and Weierstrass
and Kronecker and Cantor: I am addressing myself to the Messrs.

Waldburger and Ingelfinger, to Ms. Klubsmond and Ms. Ackeroyd, to Hafez the Intelligent, to a housewife from Santa Clara, to the frizzy-haired daughter of a French friend, unburdening herself of an adamantly expressed dislike of mathematics, and, of course, to *you,* so that like the Ancient Mariner I seem to be willing to fasten on absolutely anyone, saying with a mad look in my glittering eye—*listen, listen, listen.*

The Thunder's Roll

The fundamental theorem of the calculus has two parts, and each, in its own way, serves to unify a set of far-flung and diverse concepts. The theorem is a great *synthetic* declaration confirming what every-one is by now in a position tentatively to sense: that the concepts of the calculus are connected and connected in ways that are not trivial.

The theater master of the theorem begins with a simple, a straight-forward, assumption, one inviting a suspension of disbelief. Let f be a continuous and real-valued function on $[a, b]$, he remarks; *ah,* but note the restrictions: a *function, continuous, real-valued,* and holding court on the *closed* interval. Such is the setting in which the theorem is staged and thus the world from within which the real world rises or is represented.

The calculus begins in a dreamy meditative appreciation of con-tinuity as a feature of experience; it is by means of integration that the meditation is made meaningful and given the character of a far-reaching *theory.* The definite integral of f represents a busy operation, something undertaken if only in the imagination: *down* go the parti-tions on an interval, *up* go those Mies van der Rohe rectangles, *in* comes the beetle-browed assessor to compute their areas, the mathe-matician arriving finally as a great and imperious intellectual Prince sending the sums onward to their appointed limit. It is useful to be re-minded that all that busyness results in the end in nothing more than a number, those cloud-capped palaces on various partitions disap-pearing into thin air when their work has been completed.

If the definite integral is numerically mummified in death, in life it takes its form as an *in*definite integral when its upper limits of integration are allowed to vary; the result is a function $G(t)$ *defined* by integration:

$$\int_a^t f = G(t),$$

the definition serving to restore the integral to a community of vibrant symbolic forms.

The function $G(t)$ arises from the integration of f and so bears to f an ancestral relationship. Absent f there is *no G*; but to say all this is yet to remain within a bright cage of constructed concepts. What comes next comes in the shape of inexorability. Part I of the fundamental theorem affirms that G is a differentiable function of t, and it affirms— and *here* the reader must listen for the thunder's roll—that the differentiation of *G returns to the world nothing less than the function f itself*. The precise expression of the theorem now acquires the density that is characteristic of mathematical symbolism. If f is a real-valued function, one that is continuous on $[a, b]$, then

$$\frac{d\int_a^t f}{dt} = f(t).$$

Differentiation and integration are *inverse:* one undoes what the other achieves and vice versa. What at first appeared to be quite different mental motions acquire under the influence of the fundamental theorem the sinuous aspect of a single liquid flow, one moving like a reversible current now in one direction and now in the other. The full force of this remarkable fact can only be understood against an antecedent assessment of its improbability. Addition and subtraction are inverse to one another, but subtraction is defined in *terms* of addition, and whatever inversion results is a matter owing entirely to the definition. The fundamental theorem enforces a connection between integration and differentiation, conceived now as autonomous operations on functions, things that the mathematician (or the reader) *does*. Integration and differentiation are *independent;* each has its own distinctive feel and nature. Although both rely on the ubiquitous instrument of a limit, they belong to widely separated parts of the mathematical experience, and yet they are intimately related, the effect as strange and as wonderful as the discovery that the butterfly and the caterpillar against every expectation share a common identity. Integration and differentiation combine in a seamless mathematical figure, where one function gives rise to another that in turn gives rise to the first, and even those for whom mathematics is an affliction are often in a position to signify by an explosive puff of released air peculiar to mathematics that *this* was the goal hinted at for so long and hidden from sight no longer.

What the Thunder Says

The indefinite integral of a function f, when differentiated, returns to the place from where it started at f. This is what Part I of the fundamental theorem of the calculus says; but what, then, does it mean? And if I have asked this question before, often in the same words, it is only because the statement of a mathematical theorem suggests what it does not say and so like any other cultural artifact comes to life only in the reflective attention of its interpreters.

Imagine for a moment the landscape of the calculus *without* the theorem, so that everything is dark clouds, rain, and a gray swirling mist. Strange forms fill the night: changes in position, changes in time, difference quotients and average speed, derivatives and antiderivatives, great looming things that have area, the integrals and indefinite integrals; and each flash in the night sky seems to illuminate a shifting and confused scene with concepts continually rearranging themselves in response to the felt but unseen pressure of various powerful theorems erupting like volcanoes below the surface of the earth. To this dark Boschian landscape, the fundamental theorem of the calculus brings light, the effulgence brought about by a dramatic emptying of the conceptual arena of everything but the essential, the absolutely fundamental, instruments of analysis—those *functions* that stand poised and brooding over all, and it is precisely by means of such a drastic conceptual evacuation that the theorem imposes order on the conceptual universe.

The reciprocal relationship between differentiation and integration brings about a sense of closure among concepts, one that can be appreciated by tracking speed through its various incarnations. There is, in the first place, *position,* a sense of place, the position function indicating where an object has been and so revealing how far it has come. Differences in position, assessed against time, yield speed, a measure of how fast an object is going, its course of conduct in the world. Let speed be expressed as a function and the result is instantaneous velocity, the first of the diamond-bright concepts of the calculus, raw speed undergoing a transmutation into the derivative of position. But then the door swings wide to reveal a Cartesian coordinate system, and there speed finds itself inscribed as a curve in space, the graph of the velocity function straggling or surging across a flat expanse. The definite integral records the area underneath this curve, and the indefinite integral recasts this area as a moving target, something swelling or shrinking with time.

These motions, although exotic, take place at the threshold of the fundamental theorem. It is by means of the theorem itself that the indefinite integral of velocity brilliantly reappears as a differentiable function in its own right, one returning ineluctably to velocity. But *that* function has an identity that is already fixed, already frozen. The indefinite integral of velocity is simply and none other than the function measuring position, the function that determines the place of an object in the world.

A system of scattered impulses now coheres to form a conceptual circle. The derivative of position is velocity, a measure of how fast an object is going. The indefinite integral of velocity is thus distance, a measure of how far an object has been going fast. And the derivative of distance is again velocity, a measure of how fast an object has gone far. These relationships emerge, they are seen clearly, only when the symbolic forms with which they are expressed exist as mathematical functions, so that the fundamental theorem has the effect of *revealing* a series of connections that might otherwise be obscure. The action of the theorem and the range of its influence move from the particular to the general. It illuminates the connection between position and area and between speed and distance, and in this the theorem makes a claim that is almost directly connected to the harsh but familiar world in which motorcyclists roar down desert highways; but it also illuminates the connection between the continuous functions that represent position and area and speed and distance, and in this the theorem embodies a discovery about the very nature of continuity itself.

Differentiation and integration in their most general aspects reflect waves of intellectual contraction and expansion corresponding to two fundamentally different systems of description. The derivative of a function concentrates the mind at a point; the landscape it reveals is local. The integral of a function allows the mind to contemplate a region of space; the landscape over which it lingers is global. And this is what Part I of the fundamental theorem reveals about the inner nature of continuity: from a continuous global description, a local system may be recovered, and from a continuous local description, a global system.

In some ways, this remarkable exchange between global and local descriptions is the fundamental maneuver by which we come to understand reality, the very dance of life itself. Each of us occupies the center of something like a zone or region within which our consciousness glows like a steady red light. Where the region has a center, there *we* are; beyond are the other red lights, glowing on their own like lights seen at sea. The world impresses itself upon us at the single hot, charged center of consciousness. But the world is something that *transcends* the limits of my consciousness. The simplest of all epistemological schemes seems to enforce a distinction between what is local and particular and what is global and general.

And what is this distinction but a *coarse* and *vulgar* description of differentiation and integration? The miracle of the calculus is that in the realm of the real numbers, the passage from local to global and back again is both possible and necessary, so that whenever differentiation reveals a white-hot collection of local points, places glowing in their full particularity, integration recovers a global picture, a panorama. And whenever integration produces a panorama, a region-wide characterization, there is always a countervailing process by which those white-hot local points may be recovered and thus discovered anew.

Reflections from the Book of Nature

The calculus is a mathematical theory, a set of connected concepts, but it emerges in human history as the expression of a fantastic and unprecedented ambition, which is nothing less than to represent or re-create the real world in terms of the real numbers. Yet even in the darkest of dark ages, it is worthwhile to recall, when Europe was covered with frost and forests extended from the Channel to the Asiatic steppes and gray mists rose up out of the bogs and fens, men and women shrewdly observed the natural world, adjusting the rhythms of their lives to the great periods of the sun and the sky. They reckoned and they calculated, drawing inferences from the facts manifest to their senses and testing those inferences against the harsh and bitter judgments of experience. Call the system that results *animal empiricism,* the *animal* serving to denote the unforced character of its conclusions. It is this old, this settled way of looking out at things that the calculus rejects.

Animal empiricism describes the world as it appears to creatures content to frame theories answering to the obvious. The world it reveals is one of individuals and their properties, primitive relationships, things, events; it is the achingly human world, the one that we describe and cherish, the world in which what is real is what is familiar and what is familiar is real. It is a world coordinated by *qualitative* relationships: objects move slowly or not, things fall, the sun curves across the afternoon sky, spaces extend for a way, time passes. The theories that it commends are plainspoken, a compilation of clichés and their generalizations.

In place of those theories, the calculus holds a cold, a glaring, mirror to the face of nature. The relationships it commends—the secret network of nerves that it surveys—are *quantitative* relationships between real numbers. The characteristic fissure in modern sensibility between the world as it seems to the senses and the world as it seems to the sciences opens like a molten crack at the moment the first formula of the calculus is scribbled onto paper. The world that the calculus reflects cannot be discerned by the senses, and the world that can be discerned by the senses this mirror does not reflect.

Beyond any of the conclusions it reaches, the calculus offers an extraordinary account of a new way of representing space and time, and it is in this that it achieves its most radical, its most disjunctive break with the way that things were, the comfortable, old, habitual world.

The system of description enforced by animal empiricism applies common sense to common things. We human beings see only surfaces. An intelligence that could see directly into nature—what would it say to suggest what it had seen? Perhaps it would return nothing more than a list, as we now describe the contents of a room. *This,* that intelligence would say, is *that,* inquiry itself returning to the ancient act of naming. The great, the central affirmations of the calculus, by way of contrast, express neither the clichés of the commonplace nor the names of the ineffable. They are equations instead. They take the form: *there is some unknown function x and its derivative is f.* And this surely should appear puzzling. Why *are* the expressions of the calculus expressed as equations, why, that is, must mathematics have recourse to an infinitely infuriating, infinitely indirect system of descriptions, one that cannot be used until clues are cracked and constraints deciphered?

The answer represents the great inadvertent discovery of the calculus. It has, that answer, all of the bluntness of the inescapable. These queer symbolic forms have entered the world of thought because they *must:* no simpler system of description suffices to describe the world's network of mathematical nerves; human beings do not have direct access to things in themselves.

If the affirmations of the calculus are equations, the calculus also provides an account of the method by which the identity of their unknowns is fixed. The most general procedure returning a function from its derivative is antidifferentiation, as when $f(t) = t^3/3$ is given as an antiderivative of t^2. The equation $dx/dt = t^2$ is thus by antidifferentiation relieved of its unknown, its single clue resolving into a brisk identification. The unknown function x is $f(t) = t^3/3$. But the ease with which this example is forthcoming displaces but does not dispel the intellectual fog that has accumulated over the operation. Antidifferentiation is given in the calculus as an exercise in definition, a verbal play. Given a function f its antiderivative is *said* to be some other function F such that differentiating F returns the mathematician to *f*. The emphasis is on the *saying*. But what establishes, what procedure guarantees, that in the calculus any function at all *answers* to the definition of antidifferentiation? *Nothing yet* is the blunt response; and with *nothing yet* entered as an answer, the calculus would seem to be in the position of the sort of theology in which a great many confident assertions are made about the Deity, with every effort to determine whether there *is* a Deity returning from the void unverified and unverifiable. It is entirely possible, given the equation $dx/dt = t^2$, or any like it, that *nothing whatsoever* answers to the equation's constraints, so that these concise and pregnant affirmations conceal not so much an unknown as an awful emptiness, the equations of the calculus lined up stolidly, one behind the other, suggesting in their sterility a soundless scream.

It is the fundamental theorem of the calculus that establishes, *it declares,* that the indefinite integral of a function is one among its antiderivatives. If *f* is continuous, there is some function *G* that satisfies the equation, that makes it true. What is more, *G* has a determinate interpretation as area up to a point, and a determinate identity as an antiderivative of *f*. There is no assurance that the indefinite integral of *f* will in turn be an elementary function; but integration is a technique for creating functions, a way of bringing them into the world, so even if the antiderivative of *f* is not elementary, the fundamental theorem of the calculus establishes, *it guarantees,* that it exists as a part of the mathematician's rich symbolic apparatus.

The calculus is thus a door opened by a miracle; but the most complex, the most profound, of physical theories bear faithfully some trace of its simple differential equations, and some trace as well of its overall architecture. Whatever the theories, the pattern is always the same. A series of equations serves to describe some aspect of the physical world obliquely, by means of mathematical constraints, so that the most basic statements of the theory are couched in a language of whispers and hints suggesting, but often not stating, what answers to the equation's unknowns. In every theory, two separated methods or procedures, dealing with intellectually distinct and widely separated kinds of experience, are by an act of prestidigitation shown to be coordinated. And in every theory, something like the fundamental theorem of the calculus shows the act of pres- tidigitation to be justified.

The idea that the calculus brings to consciousness a new method of description infuses a number of varied metaphors with meaning. The determination of Galileo's law of falling objects begins with the *fact* that their acceleration is always constant. Absent the fact, there is no law. It is not the business—*it is beyond the power*—of mathe- matics to determine the acceleration of falling objects. Mathematics enables the mathematician to exploit this fact. This may seem to rel- egate mathematics to an ancillary role. Not so. It is the calculus that isolates the secret nerve between acceleration and the function that represents the position of a falling object; and it is, above all, the fun- damental theorem of the calculus that *invigorates* the nerve by show- ing that a vital current of living energy flows from the fact to the function.

The reflection of the real world that mathematics returns is abun- dant, lush beyond need. There is a sense in which the calculus does not offer a representation of this, the very particular, singular real world at all, but a representation rather of every possible world, those that are imaginary as well as the real, so that the reflection that it pro- vides, it provides by an act of imaginative carelessness, those equa- tions yielding only a hint of the real world, as when sailors guess at a coast by the perfume of its fruits smelled for the first time far out at sea.

Nonetheless, this image of abundance suggests finally the full significance of the fundamental theorem, whose deepest purpose is not to reveal the laws of nature, but to allow the laws of nature to be re- vealed.

And in this profound sense, it is the fundamental theorem of the calculus that establishes not that modern mathematical science is true, but that it is *possible*.

The Eternal Coachman

The fundamental theorem of the calculus was known to Newton and known again to Leibnitz, who with their luminous powers were able to see into the heart of darkness quite before any of the critical ideas ever were put in modern form or were made precise by mathematicians such as Cauchy. The story of the theorem is a prophetic saga. One part of the story is of special poignancy. Newton's teacher at Cambridge, Isaac Barrow, evidently understood the relationship between differentiation and integration, and he understood it before anyone else.

A clumsy but endearing color etching in a used calculus book shows Barrow as a merry-looking young man. He has a calm, even, and symmetrical brow, but the purity of his features collapses at his broad fleshy nose and his underslung jaw, which tends to push up his lower lip; it is nonetheless a gentle and appealing face, and in the entire history of mathematics it is the *only* face suggesting that the man who inhabits it might when confronted with a difficult problem confess that *he* hadn't the slightest idea of its solution. It is the relaxed face of a man prepared to be amused, and yet Barrow was a fine mathematician, no doubt one of those individuals marked from early life by his easygoing and general intelligence. There he is in the middle of the seventeenth century, a professor of Cambridge, a man of certain parts, a considerable linguist and a passionate geometer, a *figure* in academic life. And then Newton entered his world of work. Anyone of Barrow's sensitivity, I am sure, would have realized that the pulls and tugs of his own interests and associations were from the first threatened by the massive gravitational influence exerted by Newton's intelligence itself. Imagine looking out over a roomful of adolescents, pimply young men wearing starched clerical collars, and seeing Newton's coal black eyes staring back at you from the middle distance!

With a fine sense that he was doing a fine thing, Barrow surrendered the Lucasian professorship to Newton and retired from mathematics to pursue theology. Contemporaries remarked on his good grace in acknowledging Newton's genius, but really what else could he do? He was vastly too capable to ignore genius and not capable enough to rival it.

And here is the hopelessly poignant point. Barrow knew the essential aspect of the fundamental theorem of the calculus; he must have known as well that when men would come to mention the great theorem and the mathematicians who first noticed it, they would recognize handsomely the role played by Newton and by Leibnitz, and yet with just a touch of asperity in their collective voice, they would be bound in the tide of time to ask one another just who *is* this fellow Barrows?

The Great Shortcut

The first part of the fundamental theorem establishes the nature of an intellectual movement; it specifies the waves of differentiation and integration. The second part of the theorem has a different character. It shows the mathematician how to compute an elusive number easily; it is an astonishing computational tool, the first of the great algorithms to go beyond algebra and the world of algebra.

Part I of the fundamental theorem has emptied the landscape of everything save functions; with the second part of the theorem, canceled or forgotten ideas creep back into the scheme of things in order to restore their claim to conceptual legitimacy. There is the *definite* integral, to take a case in point, an item slighted by the fundamental theorem in favor of the *in*definite integral, the regnant function. Whatever its definition, the definite integral has at least the merit of returning a number to the table; but the curious and troublesome fact is that so far the number may be reached only by means of the definition of the definite integral itself, the process at once arduous and frustratingly inefficient.

Part II of the fundamental theorem shows that there is a redemptive connection between the definite and the indefinite integral; with the connection established, a technique is created by which *definite* integrals may lightheartedly be computed. If this is to return the calculus to a larger landscape than one occupied entirely by functions, it is yet to see the functions for the commanding artillery pieces that they are; the redemptive connection between definite and indefinite integrals comes about only *because* the appropriate functions have already been brought into existence.

The definite integral of a real-valued function

$$\int_a^b f$$

stands between its limits of integration where, if f is positive, it is engaged in the useful work of denoting the area underneath a curve; but whatever the work, the effort ends in a *number,* and this fact represents one of the points at which the calculus returns to its roots in the simplest of all mathematical objects.

The indefinite integral, on the other hand,

$$\int_a^t f = G(t),$$

is a function and so ferries numbers to numbers. But if f is continuous on $[a, b]$, the second part of the fundamental theorem affirms, the definite integral between a and b may always and forever be represented as the difference between $F(a)$ and $F(b)$, where F is *any* antiderivative of f.

The theorem's stress is on *any;* it is, of course, trivially true that

$$\int_a^b f = G(b) - G(a),$$

inasmuch as G was dutifully *defined* in terms of integration itself. Subtracting $G(a)$ from $G(b)$ could leave nothing but the very area measured by the definite integral. The fundamental theorem goes beyond G to embrace *any* antiderivative F; and the theorem affirms that what holds for G holds as well for F:

$$\int_a^b f = F(b) - F(a).$$

This is a powerful and productive result. To one side of the conceptual universe there is an immensely complicated series of sums, a *global* configuration extending over the whole of a region of space; to the other side of that same universe, there are two points, bright isolated pinpricks of light. The fundamental theorem asserts, *against every reasonable expectation,* that from these points of light all that is relevant to the global expanse may be deduced.

If this seems rather like an excessively fast rabbit pulled from an exceedingly threadbare hat, some reassuring sense of the theorem's power comes about when it is actually put to good use. The function $f(t) = t^3$ provides a familiar example. Efforts to calculate the area underneath the graph of this function from 0 to 1 involved a vexing series of computations, all of them devoted somehow to assessing a particular limit, *this* one, in fact:

$$\lim_{n \to \infty} \sum_{i=1}^{n} f(t_i) \Delta_i.$$

Those sums, that limit, those vexing calculations turn out to be irrelevant and unnecessary. All that is needed to determine the area underneath this curve is an acquaintance with an antiderivative of f, with *any* antiderivative of f. Easily enough done. The function

$$F(t) = \frac{t^4}{4}$$

when differentiated yields $f(t) = t^3$. This function, the fundamental theorem of the calculus asserts, may be assessed at 1 and again at 0 in order to determine the area underneath the curve. With the assessment complete, the result is $1/4 - 0 = 1/4$. And that is astonishingly enough the answer delivered before, delivered this time by means of a breezy one-line calculation.

Anyone with a sense of the world's harsh justice must be inclined to suspect that this is too good to be true.

And yet it *is* true.

This calculation, and the great theorem that makes it possible, illustrate, if only in a small way, a theme sounded in the calculus and then sounded again throughout mathematical science. The fundamental theorem says that a calculation undertaken at only two points suffices to determine the area underneath the curve. The requisite contrast is between area, which is something that extends over an entire region of space, and those two singular points, gleaming in an otherwise blank and empty sky, the isolated and detached points containing somehow information sufficient to characterize an entire region of the plane, to control the nature of that enigmatic stuff that lies beneath a curve.

Number to Number

Part I of the fundamental theorem draws a connection between *functions,* Part II of the theorem, a connection between *numbers;* but like faraway spotlights sweeping a deserted coast, the two halves of the theorem inform one another.

A falling stone, Galileo's law affirms, travels at a speed that is proportional to the elapsed time; its velocity is thirty-two feet per second, so that $\mathbf{vel}(t) = 32t$. The indefinite integral of $\mathbf{vel}(t) = 32t$, namely

$$\int_a^t 32t,$$

describes a differentiable function of t. This, Part I of the fundamental theorem reveals, is the position function $P(t) = 16t^2$. It is this function that creates a living connection between the position of a falling object as it descends downward through the region of a Cartesian coordinate system and the time during which it is falling.[1]

The position function makes possible the recovery of distance— a measure that goes beyond *where* an object has landed to specify *how far* it has actually gone. Traveling for just three seconds, a falling object consumes $16 \times 9 = 144$ feet.

Only a short series of steps is needed to cover the ground between position as a function and distance as a number. The mathematician takes these steps on the right side of the conceptual field that Part II of the fundamental theorem divides. On the left side of the field, area is under assessment, specifically, the area under the curve bounded

---

[1] For the details, see chapter 12.

by the lower edge of a Cartesian coordinate system and the graph of the velocity function between the origin and 3. The *definite* integral of velocity

$$\int_0^3 32(t)$$

denotes the required *number*. But the varied calculations that the integral would otherwise require can be carried out instantaneously by investigating any antiderivative of the velocity function between the origin and 3. Such is the burden of Part II of the fundamental theorem. Position is defined as an antiderivative of velocity. The appropriate *calculation* is simply one in which $P(0)$ is subtracted from $P(3)$. This is precisely the calculation undertaken to determine distance. The area underneath this curve is thus denoted by 144.

However strange at first, the exercise is simple enough, but like a polished glass surface from which the light bounces and winks, the example hides in a display of light that which it is meant to reflect. Applied to distance, speed, and area, the fundamental theorem of the calculus demonstrates that *as functions* these concepts are coordinated. But the calculation just completed finally ends in a particular number—144; it is a number that represents the distance traveled by a falling object and represents *as well* the area underneath a curve. One number thus has come to stand in for two quite different extensions, denoting both distance *and* area.

In one sense, this is no surprise. After all, day-to-day distance is defined as the product of speed and time, and what is the integral but a fancy version of that product? But in quite another sense, the identification of distance and area should occasion astonishment. Distance is, after all, a one-dimensional concept, the measure of something from here to there, and area is a two-dimensional concept, something bound up with territory or extent. Distance results from a simple mathematical operation. The distance between numbers is their difference. Area requires an elaborate mathematical construction before it transcends its origins in the formulas of Euclidean geometry. And yet distance and area are connected, if only by the numbers by which they are designated.

In this sense, the fundamental theorem of the calculus demonstrates a blurring of the line between what is purely mathematical and what is purely physical. Differentiation begins with a physical fact: the displacement in position of a physical body. It ends with a mathematical object: the derivative of a real-valued function, introduced in order to express instantaneous speed. Integration begins with a mathematical object: the area underneath a curve. It ends with a physical fact: the displacement in position of a physical body. This pattern and the subtle destabilization of established categories that it brings about reflect the work of the calculus at the deepest level.

The Fundamental Theorem

Part I

The proof of the fundamental theorem of the calculus is child's play. The full force of the theorem resides in what it says and not in how it is deduced; the argument requires that only a few facts be kept resident.

It is the function G about which the fundamental theorem makes large claims; and in view of the shape those claims are apt to take, it is well to have its *difference* quotient on hand from the first:[3]

$$\frac{G(t+h) - G(t)}{h}.$$

This suggests the strategy of the argument to come; both $G(t+h)$ and $G(t)$ have a direct interpretation in terms of integration and thus, if f is positive, in terms of area. My argument proceeds as *if f* were positive, and I talk of area as *if* it were a given; but nothing—*absolutely nothing*—that I say depends on this assumption, and it appears in what follows as a lighthearted concession to intuition.

The relevant inferences now tumble out one after the other. The area underneath the curve between the points a and $t+h$ is described simply as:

---

[3] In what follows, I have allowed the letter h to stand for Δt—this purely for typographical convenience and ease of presentation.

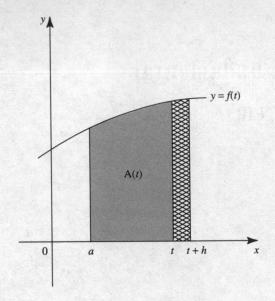

$$\int_a^{t+h} f.$$

In view of the definition of G,

$$\int_a^{t+h} f = G(t+h).$$

The area underneath the curve from a to t is described as

$$\int_a^t f.$$

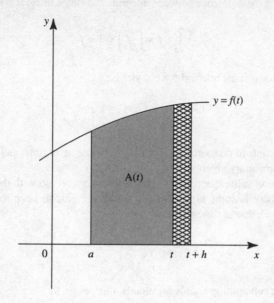

But again by virtue of the definition of G,

$$\int_a^t f = G(t).$$

The difference between $G(t + h)$ and $G(t)$ is expressed by lining up the symbols:

$$G(t + h) - G(t) = \int_a^{t+h} f - \int_a^t f;$$

and this expression effectively expresses the area underneath the curve of f in a small strip whose base is h. This might suggest, the symbols seductively *inviting* this interpretation, that in the limit the left side of this equation will see the difference become a derivative, even as the right side of the equation somehow reappears as $f(t)$ itself.

Dividing both sides of the equation by h is the first step:

$$\frac{G(t + h) - G(t)}{h} = \frac{1}{h}\left(\int_a^{t+h} f - \int_a^t f\right).$$

And recasting the difference between integrals as a single integral is the second:

$$\left(\int_a^{t+h} f - \int_a^t f \right) = \int_t^{t+h} f.$$

Substituting new integral for old yields:

$$\frac{G(t+h)-G(t)}{h} = \frac{1}{h} \int_t^{t+h} f.$$

All this amounts to no more than the play between definitions and a few elements of elementary algebra.

The mean value theorem for integrals was conceived with this moment in mind. There is some argument of *f*—call it *c* just to keep the symbols straight—such that at *c* the value of *f* is

$$f(c) = \frac{1}{h} \int_t^{t+h} f.$$

But on substituting equals for equals, this comes to

$$\frac{G(t+h)-G(t)}{h} = f(c),$$

the form of the formula now sufficient, I hope, to prompt prodromal tingling along the spines of those sensitive to symbolism.

And now a final fact falls into place, bringing the theorem to a close. Among the assumptions placed on *f* is the requirement that *f* be continuous. It follows from the nature of continuity, the implication rumbling up from the heart of things itself, that in the limit *f(c)* approaches *f(t)*. This means that

$$\lim_{h \to 0} \frac{G(t+h)-G(t)}{h} = f(t).$$

But

$$\lim_{h \to 0} \frac{G(t+h)-G(t)}{h}$$

is the derivative of *G* and this, the theorem now reveals, is none other than *f* itself.

Part II

The proof of the second part of the fundamental theorem is very easy and moves very quickly. The function G is by definition the indefinite integral of f:

$$G(t) = \int_a^t f.$$

Now, G has already established itself as an antiderivative of f. Let F be *any other* antiderivative of f so that the derivative of F and the derivative of G both coincide in f. The mean value theorem making its usual deft appearance, it follows that F and G differ only by a constant:

$$G = F + C.$$

But $G(a)$ *must be* 0. After all, G is defined in terms of the integral from a to t, and if $t = a$ itself, there is no area to compute, nothing to add, and that integral is going nowhere at all. Thus,

$$0 = G(a) = F(a) + C.$$

But this means that C is itself $-F(a)$, so that whatever t happens to be

$$G(t) = F(t) - F(a).$$

In particular, when G is evaluated at $t = b$, the very place that definite and indefinite integral for a moment coincide,

$$G(b) = \int_a^t f = \int_a^b f = F(b) - F(a).$$

The proof is in a flash complete.

chapter 26

A
Farewell
to
Continuity

THIS THEN IS MY STORY AND LIKE ALL STORIES THIS ONE CAN DO NO MORE than enclose the reader in a circle of human voices.

The calculus is humanity's great meditation on the theme of continuity, its first and most audacious attempt to represent the world, or to create it, by means of symbolic forms that in their power go *beyond* the usual hopelessly limited descriptions that we habitually employ. There is more to the calculus than the fundamental theorem and more to mathematics than the calculus. And yet the calculus has a singular power to command the attention of educated men and women. It carries with it the innocence of an abstract pursuit successfully accomplished. It is a great and powerful theory arising at the very moment human beings contemplated the infinite for the first time: sequences without end, infinite additions, limits flickering in the far distance. There is nothing in our experience that suggests that mathematics such as this should work, so that the successes of the calculus in unifying aspects of experience are tantalizing but incomplete evidence that of the doors of perception, some at least may open and some at least may lead to someplace beyond.

Having made modern mathematical science possible, no doubt the calculus made it inevitable as well. No purely physical theory has ever severed its link to the calculus nor severed its reliance on the prestidigitation that the calculus requires and embodies. But for all the power and real intellectual grandeur of contemporary scientific schemes, involved as they are in the description of strings or the cosmic inflation that took the universe from a bang to a bubble in the twinkling of an eye, the enterprise of which they are the supreme expression no longer commands wide assent as a secular faith. I say this as no mark of disrespect. It is simply a fact. There is a fissure in contemporary thought, physicists arguing that each advance brings them closer to a final theory and the rest of us observing that the difference between what has been and what needs to be accomplished remains what it has always been, which is to say infinite. The simple melancholy fact is that outside the charmed circle of those working on the current frontiers, no one believes any longer that physics or *anything like physics* is apt to provide contemplative human beings with a theoretical arch sustaining enough to provide a coherent system of thought and feeling.

And yet human beings are a naturally inquisitive species, and if the questions we would ask at the very far margins of our experience—*how did it begin and in what and why?*—have even in the asking a hollow and self-mocking quality, as if the universe were designed to discourage such speculation, there are plenty of other questions that provoke our curiosity; and the withdrawal from the grand concerns of physical theory may well indicate as much a change in attitude and interest as an intellectual defeat.

Biologists, for example, appear to possess what physicists now lack: a commonly agreed upon method, an accepted intellectual agenda, and a set of research problems accessible both in economic and intellectual terms. This would occasion no more than a shrug were it not for the strange fact that molecular biology is so very different a discipline than anyone might have expected. No mathematics, for one thing. Despite a few attempts by mathematicians here and there to participate in the life of the biological sciences, mathematics has played *no* role in molecular biology and seems destined to play none. No achievement in molecular biology requires mathematics beyond finger counting for its comprehension. But even stranger, there is this: that the thought world of molecular biology would in its major aspects be instantly comprehensible to someone who knew nothing of science, modern physics, Newton, continuity, or the calculus. Living systems may best be understood in terms of their constituents. Going down, one encounters organ systems, organs, tissues, cells, cell parts, and then on a much smaller scale of organization, molecular constituents of which the most important are the proteins and a master molecule, DNA. But *there,* in contrast to physics, things come to an end. In place of depth, the biologist requires intellectual extent. He or she wishes to trace connections among the biological constituents, following pathways across a living system and coming to understand how influences are transmitted.

This is an oversimplification only in the sense that it is the details that need to be filled in. The outline is clear enough. It reveals an intellectual landscape far simpler than the one inhabited by mathematicians. Mathematical science requires *theories,* molecular biology, *facts.* As one century gives way and another comes to replace it, the very nature of science as a distinctive human activity is ineluctably changing.

The contrast between the mathematician and the biologist is one drawn in terms of two different intellectual attitudes, two different strategies for confronting experience. In one, adequacy of description is traded for *depth* of insight, and this is the strategy chosen by modern science and by Western philosophy. It is the strategy that receives a supreme expression in the calculus, for everywhere in the calculus there is a ruthless rejection of the clutter of experience in favor of a world re-created in terms of real numbers and functions of real numbers. The merit of this way of proceeding is that it reveals the essentials; its defect is that it slights the character of experience. Theories may revisit the facts, as when they make successful predictions; they may, indeed, they *do,* function in a vital way in the manipulation of nature on a small scale, as when mathematics is *applied;* they may have an overwhelming intellectual authority; but they are not, *they cannot be,* adequate to the character of experience as it is recorded in ordinary life, adequate, that is, to the thousand shimmering if evanescent connections that exist between one person and another, between one place and another, and between one time and another.

To say what mathematical science cannot do is promptly to redeem a second intellectual strategy, one in which depth is traded for adequacy of description. The aim of such a strategy is not to re-create the world but to describe it. Its origins lie in the immemorial animal empiricism that informs our unforced and natural account of the real world. It is the strategy that receives a supreme expression in modern biology, for everywhere in biology there is an indifference to ultimate causes and irreducible constituents—no biologist would think of explaining the metabolism of a bat in terms of quarks—and in place of this concern a passionate curiosity about connections, patterns of influence, the ways in which a biological system *works*.

It may well be that human beings are, by virtue of the way in which they have been made, partial to biological explanations, inclining instinctively to the accumulation of facts and the solid and comforting sense that they convey of dealing in the details that count. In this sense, modern molecular biology continues an ancient tradition.

Beyond biology, what? We all live within a dense and reticulated network of connections and causes, contingencies and correspondences, a network that is itself alive and quivering with human passion and sensibility; it is that web of dependencies into which we are born and that web from which we depart when we die. An account of that web, an instantaneous and accessible sense available to every one of its members, would have little to do with modern mathematical science, nothing in its origins suggesting the calculus. It would be an account almost entirely of appearances, of how things in their multifarious ways are coordinated and connected: it would be a theory overwhelmingly of facts, of things as they are given to us in the here and now where we live and breathe and pass the time. The dream of understanding things in this dynamic way has been an intermittent part of human intellectual aspirations since time immemorial. It was in part a dream dreamt by Leibnitz himself, whose strange lucid genius now comes to loom over the late twentieth century. Why surrender the world of appearance, he might ask, if the world of appearance may be *completely* understood? But it has been only within the last half century that human beings have in the computer an instrument capable at least in principle of dealing with the sheer size of the web, its complexity. And this, too, Leibnitz foretold as well.

The computer cannot think; it cannot act; it has no volition or purpose; but there is an eerie economy of effect in its operations and a genius of a singular order in its design, a form of cunning commanded by no other intellectual instrument. It achieves its striking results by simplifying its organization so that it encompasses a few basic logical operations. It addresses a world presented in chunks of data and by virtue of its great speed and simplicity manages to coordinate aspects of that world *directly,* with no mediation of theory, no appeal to abstract concepts. The computer maintains no contact with the concepts of continuity. It is supremely an instrument by which connections are tracked in time and then recorded. If the calculus embodies, or at least represents, an ancient human urge toward theoretical abstraction, the computer represents, and may embody, an equally ancient human urge toward factual mastery.

Whither continuity in all this? The long and extraordinary meditation on its meaning is coming to an end. The mathematics that has gone into the meditation has become too rebarbative and the system of rules by which it is animated too complicated to sustain a large community of purpose. It requires unusual abilities to become a mathematician, that and years of painful training in which the intellect is forced to bend upon itself. Like sixteenth-century counterpoint, or the rituals of the Persian Court, the thing has become overly elaborate, and in science as in art what is overly elaborate is destined to disappear.

For those of us who like the Persian courtier have grown to accept the complexities of mathematics and have allowed the bizarre and very difficult to become familiar, there is a natural tendency to mistake the world we inhabit for the world at large, and as the courtier cannot imagine life outside the great and stately Court, with its palm trees, the smell of incense and frangipani, and the deep thrilling purple of the imperial insignia, so the mathematician cannot imagine forms of intellectual experience that are not in some sense dominated by the ancient ideas of the continuum and its properties and powers.

Yet everything has a beginning, everything comes to an end, and if the universe actually began in some dense explosion, thus creating time and space, so time and space are themselves destined to disappear, the measure vanishing with the measured, until with another ripple running through the primordial quantum field, something new arises from nothingness once again.

epilogue

*In the course of Time, these Extensive Maps were found some-
how wanting, and so the College of Cartographers evolved a
Map of the Empire that was of the same scale as the Empire and
that coincided with it point for point. Less Attentive to the Study
of Cartography, Succeeding Generations came to judge a Map
of such Magnitude Cumbersome, and, not without Irreverence,
they abandoned it to the Rigours of Sun and Rain. In the Western Deserts, tattered Fragments of the Map are still to be found,
sheltering an occasional Beast or Beggar; in the whole Nation,
no other relic is left of the Discipline of Geography.*

JORGE LUIS BORGES, "OF EXACTITUDE IN SCIENCE"

acknowledgments

I have written this book in isolation, hardly talking to a soul, but un-oppressed as well by campus codes or creeds, free to say what I want and when I want. It is a measure of the degradation that has overtaken American academic life that I should feel obliged to boast of such circumstances. I am grateful beyond measure to my wife, Victoria, for making my freedom possible. In everything I have done, it has been my hope to write something worthy of her admiration.

Susan Ginsburg is the world's greatest literary agent; but she is also a fine editor, at once sympathetic, discriminating, and demanding. It was Susan who saw the merits to this book at a time when it existed only as a one-paragraph proposal, and Susan who time and again insisted in her own patient but implacable manner that whatever I had done I could do better. She was right. I have through experience come to suspect that she is always right, and if the book that has resulted does not yet meet with her full approval, it is not for my want of trying.

From time to time, every writer imagines that his editors are his enemies, existing only to slash in indignant red his most treasured phrases or pet paragraphs. In my own case, the reverse has been more nearly true. Dan Frank and Marty Asher have a fine ability to spot what is best in their author's prose. They have as well a determination to purge what is crude, or clumsy, or offensive, or obscure, or vulgar and ornate. "You have again said nothing at great length, Mr. Berlinski," a college instructor in English once wrote on one of my papers; and by some queer, inexplicable division of the genetic stream, he seems, that minatory and crisply remembered figure, to have been reborn in the persons of my editors. Behind the book they read, they saw the better book I should have written, and let me know, often in no uncertain terms, how much I had to do before what I wished to say coincided with what I said.

I have been influenced in the development of my thoughts by three mathematicians:

M. P. Schutzenberger has provided me with an enduring model of the mathematical intelligence: passionate, wide-ranging, courageous, and skeptical. I measure virtually everything I write against the imagined snort of his derision. Our friendship has been the most extraordinary of my life.

It was reading René Thom's work on the singularities of smooth maps that caused the frozen sea of my own intellectual self-satisfaction to shudder and then crack. More than anything else, it has been this work that has persuaded me that philosophy without mathematics is an impoverished discipline.

When we were both young, Daniel Gallin and I collaborated on a number of mathematical projects. We rented a sunny studio in San Francisco and talked away the golden afternoons. It was those conversations that revealed to me not what mathematics was—that I thought I knew—but how it should be done. I will always treasure the memory of the days we spent together, when we thought that time would never end.

index